SPORTS MATH

An Introductory Course in the Mathematics of Sports Science and Sports Analytics

TEXTBOOKS in MATHEMATICS

Series Editors: Al Boggess and Ken Rosen

PUBLISHED TITLES

ABSTRACT ALGEBRA: AN INTERACTIVE APPROACH, SECOND EDITION
William Paulsen

ABSTRACT ALGEBRA: AN INQUIRY-BASED APPROACH
Jonathan K. Hodge, Steven Schlicker, and Ted Sundstrom

ADVANCED LINEAR ALGEBRA
Hugo Woerdeman

APPLIED ABSTRACT ALGEBRA WITH MAPLE™ AND MATLAB®, THIRD EDITION
Richard Klima, Neil Sigmon, and Ernest Stitzinger

APPLIED DIFFERENTIAL EQUATIONS: THE PRIMARY COURSE
Vladimir Dobrushkin

COMPUTATIONAL MATHEMATICS: MODELS, METHODS, AND ANALYSIS WITH MATLAB® AND MPI, SECOND EDITION
Robert E. White

DIFFERENTIAL EQUATIONS: THEORY, TECHNIQUE, AND PRACTICE, SECOND EDITION
Steven G. Krantz

DIFFERENTIAL EQUATIONS: THEORY, TECHNIQUE, AND PRACTICE WITH BOUNDARY VALUE PROBLEMS
Steven G. Krantz

DIFFERENTIAL EQUATIONS WITH MATLAB®: EXPLORATION, APPLICATIONS, AND THEORY
Mark A. McKibben and Micah D. Webster

ELEMENTARY NUMBER THEORY
James S. Kraft and Lawrence C. Washington

EXPLORING CALCULUS: LABS AND PROJECTS WITH MATHEMATICA®
Crista Arangala and Karen A. Yokley

EXPLORING LINEAR ALGEBRA: LABS AND PROJECTS WITH MATHEMATICA®
Crista Arangala

GRAPHS & DIGRAPHS, SIXTH EDITION
Gary Chartrand, Linda Lesniak, and Ping Zhang

PUBLISHED TITLES CONTINUED

INTRODUCTION TO ABSTRACT ALGEBRA, SECOND EDITION
Jonathan D. H. Smith

INTRODUCTION TO MATHEMATICAL PROOFS: A TRANSITION TO ADVANCED MATHEMATICS, SECOND EDITION
Charles E. Roberts, Jr.

INTRODUCTION TO NUMBER THEORY, SECOND EDITION
Marty Erickson, Anthony Vazzana, and David Garth

LINEAR ALGEBRA, GEOMETRY AND TRANSFORMATION
Bruce Solomon

MATHEMATICAL MODELLING WITH CASE STUDIES: USING MAPLE™ AND MATLAB®, THIRD EDITION
B. Barnes and G. R. Fulford

MATHEMATICS IN GAMES, SPORTS, AND GAMBLING–THE GAMES PEOPLE PLAY, SECOND EDITION
Ronald J. Gould

THE MATHEMATICS OF GAMES: AN INTRODUCTION TO PROBABILITY
David G. Taylor

A MATLAB® COMPANION TO COMPLEX VARIABLES
A. David Wunsch

MEASURE THEORY AND FINE PROPERTIES OF FUNCTIONS, REVISED EDITION
Lawrence C. Evans and Ronald F. Gariepy

NUMERICAL ANALYSIS FOR ENGINEERS: METHODS AND APPLICATIONS, SECOND EDITION
Bilal Ayyub and Richard H. McCuen

ORDINARY DIFFERENTIAL EQUATIONS: AN INTRODUCTION TO THE FUNDAMENTALS
Kenneth B. Howell

RISK ANALYSIS IN ENGINEERING AND ECONOMICS, SECOND EDITION
Bilal M. Ayyub

SPORTS MATH: AN INTRODUCTORY COURSE IN THE MATHEMATICS OF SPORTS SCIENCE AND SPORTS ANALYTICS
Roland B. Minton

TRANSFORMATIONAL PLANE GEOMETRY
Ronald N. Umble and Zhigang Han

TEXTBOOKS in MATHEMATICS

SPORTS MATH

An Introductory Course in the Mathematics of Sports Science and Sports Analytics

Roland B. Minton

Roanoke College

Salem, Virginia, USA

CRC Press is an imprint of the
Taylor & Francis Group an informa business

A CHAPMAN & HALL BOOK

CRC Press
Taylor & Francis Group
6000 Broken Sound Parkway NW, Suite 300
Boca Raton, FL 33487-2742

Printed on acid-free paper
Printed at CPI on sustainably sourced paper
Version Date: 20160816

International Standard Book Number-13: 978-1-4987-0626-1 (Hardback)

Visit the Taylor & Francis Web site at
http://www.taylorandfrancis.com

and the CRC Press Web site at
http://www.crcpress.com

Contents

Preface

This is a textbook for a course that does not exist. Like the myths that are busted in Chapter 3, that statement has some truth to it. As Obi Wan Kenobi would say, it is true "from a certain point of view." There are a number of existing courses with Sports Science and Sports Analytics in the titles, created by intrepid professors venturing into the unknown. The topics and emphases of such courses vary dramatically, so that there is no consensus on what a course in Sports Science and/or Sports Analytics should be.

There are conferences on sports analytics. The MIT/ESPN Sloan Sports Analytics Conferences are graced with outstanding speakers, and the demand for tickets grows exponentially. The topics at a conference can range from *Moneyball* to marketing strategies, management strategies or technological breakthroughs. Regional conferences such as the Carolina Sports Analytics Meeting provide support for the increasingly large number of faculty and students doing research in sports-related areas.

To use a golfing analogy, writing a book like this is like hitting a drive at a driving range; there are many directions you can go without going out of bounds. At the driving range, I pick out a small target to focus on, and that is what I have done here. I have chosen a sample of topics that I know something about and that I find very interesting. Ideally, users of this book will have enough to choose from to suit whichever version of a sports course is being run.

The course that I have taught at Roanoke College since 1988 is a mix of physics, physiology, mathematics, and statistics. The order (and level of emphasis) of the topics has changed over the years; this book reflects the current status of my course. It is, admittedly, an eclectic mix of topics (at the driving range, I may aim at one target, but I do tend to spray balls all over the range). I hope to provide ideas and resources to help students launch projects. An important part of my course is the term project, and I have almost always been pleasantly surprised at the quality of work done in a short period of time.

I suspect that the high quality of work is due to the students' high level of motivation; not from any talents of mine, but because many students (of both genders) find it exciting to think about sports and to complete a research agenda. Sports problems are easy to create and state, even for students who do not live sports 24/7. Sports are part of their culture and knowledge base, and the opportunity to be an expert on some area of sports is invigorating.

This should be the primary reason for the growth of sports courses: the topic provides intrinsic motivation for students to do their best work.

This, as I said, is a textbook. That fact alters the literary qualities of the writing. My intention is for it to be easy and enjoyable to read, but examples and exercises necessarily interrupt the normal flow of text. As well, the exercises guide students to some very interesting results, so that some of the best discoveries about sports may be hiding in the exercises. I encourage you to look for fun facts in the exercises.

The choice of mathematical level is problematic for a book like this. Some of my favorite results require calculus or even differential equations for a full explanation, but I do not want to narrow the audience to the mathematically advanced. I have split the difference on calculus. I am not assuming that you know calculus, but I will show you some of the things that calculus can do for you. Those of you who have taken calculus can read the "calculus box" sections in the text and work the exercises labeled as calculus exercises. If you have not taken calculus, simply navigate around those well-marked areas of the book.

The extent to which a background in probability and statistics is required is more difficult to say. Sports analytics relies heavily on sound statistical reasoning. Statistical "common sense" is assumed throughout, but the details of tests and calculations are all provided. Similarly, a familiarity with the ideas of computing is assumed, but no programming is required. The reader's experience will be greatly enhanced by frequent use of the internet, spreadsheets, and calculations.

I should admit that I like to read books; I enjoy holding physical books. On the other hand, I now buy most of my books and music in digital format. And I am slowly allowing myself to stream a movie or music online and let it slip away without claiming possession. The point of this ramble is that while I recognize that the future of sports research is digital with remote access, this book has a fairly standard format. There will be a website at www.roanoke.edu/mcsp/minton/SportsMath.html (I know, I'm showing my age by posting a url that will change. A search for "Minton Sports Math" should do it, but you don't need me to tell you that). I'll post links, references, notes, and anything else that comes to mind that could be useful and does not fit the classic book mold. Ideally, part of the site will even be wiki-like.

In the last thirty years, data collection has progressed from repeated viewings of grainy videos to nearly continuous data streaming from sensors attached to every part of an athlete, from Bill James painstakingly copying box score numbers from *The Sporting News* to a one-minute online search that lists the top fifty hitting streaks in MLB history. My hope is that this book opens up some of the astounding possibilities of sports research, while helping you learn more about the games you enjoy.

This book would not exist without the encouragement of my editor, Bob Ross, who over the years has furthered my career in multiple ways. Thanks, Bob! My Roanoke College family has provided support in several ways. Dave

Taylor has listened to countless musings and rants on all aspects of the book, and provided good counsel at all times. His assistance with the joys of TeX is invaluable. Adam Childers provided much-needed statistical backing, plus hours of enjoyable sports talk. Thanks to Karin Saoub and Chris Lee for their assistance. The athletic department, especially Ryan Pflugrad, Matt McGuire, Page Moir, Scott Allison, and Chris Kilcoyne are great to work with. An important chunk of the time to do this enjoyable work was provided by the M. Paul Capp and Constance Whitehead Endowed Chair, for which I am very grateful. Paul is a great supporter of education, especially in mathematics and physics. Thanks to Dean Richard Smith for his support; it is very cool to get to cite my Dean's publication in this book! Finally, to my wife Jan and children Kelly and Greg, who deal with me in writing mode, which is even grumpier than usual: thanks for being who you are, and for your love.

Dr. Roland Minton
MCSP Department
Roanoke College
Salem, VA 24153
minton@roanoke.edu

List of Figures

List of Tables

Chapter 1

Projectile Motion

Introduction

Basketball star Stephen Curry launches a 3-point shot. As the ball traces its high arc toward the basket, fans rise to their feet in anticipation. Will it go in? Is it a little short? Similar tension accompanies a Jordan Spieth tee shot, an Andy Murray passing shot, a long football pass by Peyton Manning or Lionel Messi, or a long fly ball by Mike Trout. We will analyze the flights of balls in this chapter as we explore the area of physics known as mechanics.

Along the way, we will answer such questions as: How does Blake Griffin hang in the air when dunking? What is the optimal angle to shoot a free throw? Why do golf balls have dimples? Does a knuckleball really dance? The answers are to be found in the fundamentals of physics.

Figuring with Newton

Sir Isaac Newton (1643-1727) constructed a framework for the analysis of objects in motion. The second of his three Laws of Motion is the launching point for most of our investigations in this chapter. The shorthand version of Newton's Second Law is

$$F = ma$$

where F is the sum of all forces acting on an object, m is the object's mass, and a is the acceleration of the object. One of the most remarkable aspects of

Newton's Second Law is that it can also be written as $\mathbf{F} = m\mathbf{a}$, where $\mathbf{F}$ and $\mathbf{a}$ appear in bold to indicate that they are multidimensional vector quantities. We will return to this form of the equation when we look at motion in two and three dimensions. The mass m is a **scalar** (real number) that is related to weight: for earthbound sports, weight is approximately equal to mass times the gravitational constant g.

To keep it simple, let's start with one-dimensional motion; vertical motion, to be precise. In this case, the object's position can be tracked by its height h above some reference point (e.g., the ground). We define **velocity** as the rate of change of position with respect to time. At a constant speed, this means that velocity equals change in height divided by change in time: $v = \dfrac{\Delta h}{\Delta t}$. This gets complicated when velocity is not constant. In general,

$$\text{Average velocity} = \frac{\Delta h}{\Delta t}$$

and, for small time intervals, (instantaneous) velocity is approximately equal to average velocity: $v \approx \dfrac{\Delta h}{\Delta t}$. With calculus, we can simply say that velocity is the derivative of height. Either way, note that v can be negative (if height is decreasing) or positive (if height is increasing). The **acceleration** a of the object is, in turn, the rate of change of velocity. Then $a \approx \dfrac{\Delta v}{\Delta t}$ and acceleration is the derivative of velocity.

Example 1.1 Suppose a ball falls from a height of 50 meters. If gravity is the only force on the ball, find the velocity of the ball after $t = 1$ second and $t = 1.5$ seconds.

Solution. For most sports situations, we can assume that the acceleration due to gravity is a constant $-g$ with $g \approx 9.8$ m/s^2 or $g \approx 32$ ft/s^2. An acceleration of 9.8 m/s^2 in the negative direction means that in every second the velocity decreases by 9.8 m/s. Assuming that the ball starts with velocity 0, then at $t = 1$ second the velocity has decreased to -9.8 m/s. In the next half-second, the velocity decreases by $0.5(9.8)$ m/s $= 4.9$ m/s. At time $t = 1.5$ s the velocity has decreased to $(-9.8 - 4.9)$ m/s $= -14.7$ m/s. The ideas from this basic example will be used again for the more complicated situation of Figure 1.9.

Speed is defined as the absolute value of velocity. In Example 1.1 above, at time $t = 1$ the ball's velocity is -9.8 m/s but its speed is 9.8 m/s (downward).

Notice that Example 1.1 did not ask for heights. Because the ball's velocity is changing, the calculation of position from velocity requires more than multiplying velocity by time. Fortunately, calculus gives us some simple formulas to use, shown below in Table 1.1.

In Example 1.1, we have $c = -g$, $v_0 = 0$, and $p_0 = 50$, so the height at time t is $-4.9t^2 + 50$ m. At $t = 1$, the ball is at height $-4.9 + 50$ m $= 45.1$ m, while at $t = 1.5$ the ball is at height $-4.9(1.5)^2 + 50$ m $= 38.975$ m.

TABLE 1.1: Formulas for Constant Acceleration

acceleration	$a = c$
velocity	$v = ct + v_0$
position	$p = \frac{1}{2}ct^2 + v_0 t + p_0$

Hangin' with MJ: 1-D Motion

Using the equations in Table 1.1, we can discover an interesting fact about vertical motion. We start with a straightforward calculation.

Example 1.2 A man jumps from the ground with an initial velocity of 16 ft/s, under the force of gravity. (a) How long does he stay in the air? (b) How high does he go?
Solution. We use Table 1.1 with $c = -32$, $v_0 = 16$, and $p_0 = 0$. (Note that gravity pulls in the negative direction, while the jump is in the positive direction.) Then velocity is $v = -32t + 16$ ft/s and position is $h = -16t^2 + 16t$ ft. Now, let's decipher the questions being asked. (a) What does "in the air" mean? He is in the air from launch time (height 0) to landing time (height 0). Both times occur at height 0, when $h = -16t^2 + 16t = 0$. So, solve this equation! If $-16t(t-1) = 0$, then $t = 0$ or $t = 1$. He launches at $t = 0$ and lands at $t = 1$, hence is in the air for 1 second. (b) At the top of a jump, velocity is 0: no longer going up, not yet coming down. This occurs when $v = -32t + 16 = 0$ or $t = \frac{1}{2}$. Now that we know *when* he reaches his peak, we can determine his height using the position function. The height at time $t = \frac{1}{2}$ is $h = -16\left(\frac{1}{2}\right)^2 + 16\left(\frac{1}{2}\right) = -4 + 8 = 4$ feet.

The solution of Example 1.2 follows a pattern that you should use in most such problems. First, get the equations of motion by filling in the constants in Table 1.1. Then, solve one of the equations for time t based on the situation (e.g., how long the object is in the air, or when it reaches its peak). Finally, substitute this time value into another equation to find the quantity of interest.

The 48-inch jump of Example 1.2 is in legendary leaper status, up there with Michael Jordan and Blake Griffin. But, why do these prodigious leapers seem to hang in the air? One reason is that *all* objects hang in the air. The graph of height versus time in Figure 1.1 and Table 1.2 below show the height for the jumper in Example 1.2 at equal quarter marks in time.

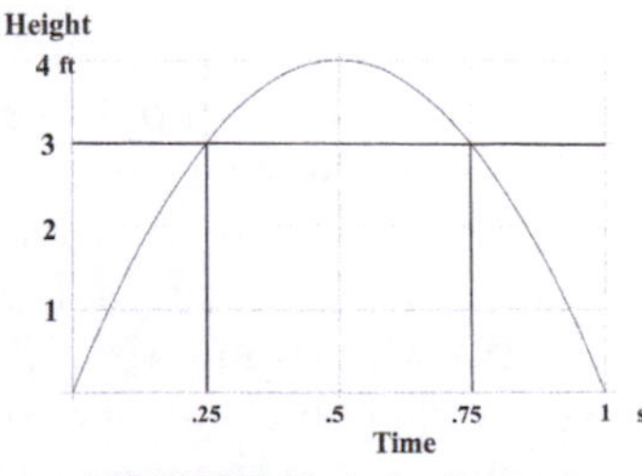

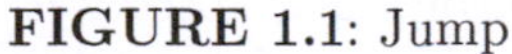
FIGURE 1.1: Jump

Notice that from time $t = 1/4$ to $t = 3/4$ (which is half of the time of the jump) the height is 3 feet or above (with a peak height of 4 feet). That is,

TABLE 1.2: Heights and Times for Jump

Time (s)	Height (ft)
0	0
1/4	3
1/2	4
3/4	3
1	0

half the time is spent in the top one-quarter of the jump! The speed is smallest at the top of the flight, so the object "hangs" at the top.

A second reason that great athletes can appear to defy gravity has to do with **center of mass**. The center of mass is where the sums of mass-times-distance quantities balance. For a standing human being, it is not far from the geometric center of the body. Newton's equations track the center of mass of the object in flight. Figure 1.1 does not show a body in flight, but the path of a single point. That point is the center of mass of the person. (Which means that a "height" of 0 does not actually mark the location of the ground; it marks the location of the center of mass of the object at launch time.) While the dunker's center of mass is tracking the nice parabola shown, he is free to pull up his legs, bob his head, and extend an arm in entertaining ways that may cause an individual body part such as the head to remain at the same height for a noticeable amount of time.

Raining 3's with Steph: 2-D Motion

Let's return to Stephen Curry's 3-point shot. We can analyze its flight with Newton's Second Law, but the fact that the ball now moves both horizontally and vertically complicates the calculations.

From nba.com/Stats, we can get an idea of the location of Curry's shot. In 2014-15, only 79 of Curry's 618 3-pointers were from the corners. (Remarkably, he made well over 40% of his shots from every 3-point zone and 62% from the left corner, plus an outrageous 91% from the left corner during the playoffs.) Most of his shots were from beyond the arc that is 23.75 feet from the basket. Let's say his shot is from 25 feet away. Align the x-axis horizontally from Curry to the basket, and the y-axis vertically.

We will assume that Curry's impeccable form keeps the ball from curving left or right. Newton's Second Law is the vector equation $\mathbf{F} = m\mathbf{a}$ where the vectors $\mathbf{F}$ and $\mathbf{a}$ have two components. That is, the acceleration has a horizontal component a_x and a vertical component a_y. Assuming that gravity

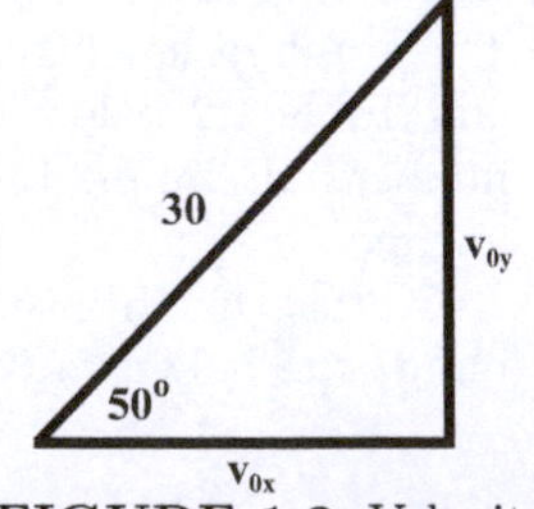

FIGURE 1.2: Velocity

is the only force, then $a_y = -g$ as before, and $a_x = 0$ (no forces acting horizontally). This allows us to separate the x- and y-equations. To use Table 1.1, we need the initial velocities and initial positions. We assume that $p_{0x} = 0$ ft for convenience and $p_{0y} = 7$ ft (assuming the ball is released from a height of 7 feet). If the ball is launched with speed 30 ft/s at an angle of 50 degrees, then v_{0x} and v_{0y} are obtained from the triangle in Figure 1.2.

Using basic trigonometry, we get initial velocities $v_{0x} = 30\cos(50°)$ ft/s and $v_{0y} = 30\sin(50°)$ ft/s, or $v_{0x} \approx 19.28$ ft/s and $v_{0y} \approx 22.98$ ft/s. Pulling this all together, we have $x \approx 19.28t$ and $y \approx -16t^2 + 22.98t + 7$.

Example 1.3 Is this shot good or not?
Solution. In this case, a perfect shot would pass through $x = 25$ and $y = 10$ (the height of the basket). We will solve for t in one equation and plug into the other equation, but that can be done in two ways. For reasons you will see, it is more convenient to start with the y-equation. We want $y = 10$ and so solve $-16t^2 + 22.98t + 7 = 10$ for t. There are two solutions, one representing the ball rising up through the height $y = 10$ and the other representing the ball dropping through the height $y = 10$; the second solution is clearly the one of interest. We get $t \approx 1.29$ s. If the shot is perfect, then at this t-value we get $x = 25$ (be sure this makes sense to you!). Instead, our equation gives $x \approx 24.90$ feet. Not perfect, but is this close enough? The center of the basket is at $x = 25$, so $x = 24.90$ represents 0.1 foot or 1.2 inches from the center. The basket has diameter 18 inches and the ball has diameter 9.5 inches, so the ball can move a little over 4 inches from the center and still be inside the basket. (This assumes that the shot is exactly on line.) Count the three!

The work in Example 1.3 does not fully prove that the shot is good. Can you think of what is missing?

Even if the center of the ball (theoretically) passes inside the basket, in real life if the trajectory of the ball is too flat some portion of the ball will hit the rim. You will show in exercise 1.41 that the ball in Example 1.3 enters the basket at an angle of about 43 degrees, more than steep enough to safely pass through the basket.

We can now develop a method to determine the best angle at which to shoot a free throw. An important part of our interpretation of the numbers in Example 1.3 is the margin of error inherent in playing with a ball that is smaller than the basket. You could imagine decreasing the initial speed from 30 ft/s until the shot is no longer good; call this speed s_1. Then find the maximum speed s_2 for which the shot is good. For the angle 50 degrees, $s_2 - s_1$ is the margin of error in speed. The bigger the margin of error, the better, since the shooter does not have to be as precise with the launch speed. Peter Brancazio has done this study and found that a free throw angle of about 49 degrees gives the largest margin of error. We will explore an interesting aspect of this angle in exercise 1.9.

K's with Kershaw: Terminal Velocity and Drag Forces

Our first three examples all utilized an assumption that gravity was the only force to be considered. Such an assumption is valid on the Moon, but is not realistic in any sports situation. So, why would we consider such examples? The only reason is mathematical convenience. While the mathematics in Example 1.3 may have become uncomfortably detailed, the underlying equations in Table 1.1 are about as simple as equations get. For a vertical jump and a basketball shot, we can hope that other forces do not have a very large effect, but for many sports this hope is in vain. In this section, we add air resistance to the mix. The good news is that the mathematical model becomes more accurate; the bad news is that the resulting equations are, for the most part, unsolvable.

Let's start with a thought experiment. If you hold your hand out of the window of a moving car, how much force will you feel? The faster the car is going, the more force, right? Also, you can increase or decrease the force by changing your hand position: more hand facing the front of the car means more force. These illustrate the main principles of **air drag**. The magnitude of the force depends on the speed of the object. In most sports situations, a good approximation is that the force due to drag is

$$F_d = cv^2$$

where F_d is the magnitude of the drag force, and v is the speed of the object. The scalar c depends on such influences as humidity, temperature, altitude, the composition and orientation of the object, and other factors. One of the other factors is the speed of the ball, but to keep things simple we will ignore this. Most of the influences are well known to us: at high altitude the air is thinner and air drag is reduced: the air drag at elevation 7350 feet (e.g., Mexico City) is 23% less than that at sea level. Under most conditions, higher temperature reduces air drag. The main reason that balls fly farther in warm weather is humidity: contrary to most people's intuition, higher humidity causes *lower* air drag.

For a falling object, Newton's Second Law now looks like

$$a = -g + kv^2$$

where k equals the c from the drag force divided by the object's mass m. Note the signs: gravity pulls in the negative direction while (since the object is falling) air drag pushes upward. Let's track acceleration in another thought experiment: if you jump out of an airplane, what will happen to your velocity? At first, kv^2 is smaller than g so that a is negative. A negative change in velocity means that your downward speed will increase. But, the faster you fall the more air drag you experience and your speed gradually approaches an equilibrium value at which the air resistance kv^2 exactly balances gravity g. At this point, your acceleration is zero and your velocity will remain constant. (This is Newton's First Law: an object maintains a constant velocity unless acted on by an external force.) You will continue to fall at this speed unless there's a change in force: e.g., your parachute opens, or you change body orientation to alter air resistance. If your parachute does not open, the name **terminal velocity** for this speed is appropriate, if harsh.

Example 1.4 If the terminal velocity of a baseball is 95 mph, find the value of k for air drag on the baseball.
Solution. One of our values for g is 32 ft/s^2, so we need to convert mph to ft/s. Given 5280 feet in a mile and 3600 seconds in an hour, we have $95\ \frac{\text{mi}}{\text{hr}} = 95\frac{\text{mi}}{\text{hr}} \cdot 5280\frac{\text{ft}}{\text{mi}} \cdot \frac{1}{3600}\frac{\text{hr}}{\text{s}} \approx 139.3\frac{\text{ft}}{\text{s}}$. At terminal velocity, $a = 0$ so $-g + kv^2 = 0$. Then $k = \dfrac{g}{v^2} = \dfrac{32}{(139.3)^2}\frac{\text{ft/s}^2}{\text{ft}^2/\text{s}^2} \approx 0.00165\ \text{ft}^{-1}$.

We can now estimate how many miles per hour a Clayton Kershaw fastball loses as it travels to home plate. To do this, we use the above value for k and assume that a pitch travels horizontally. We also assume that the pitch travels 55 feet (60 feet, 6 inches minus a long stride and stretch).

Example 1.5 For a baseball that starts at 95 mph and travels horizontally (no gravity), find its speed after it travels 55 feet.
Solution. In the absence of gravity, Newton's Second Law is $a = -0.00165v^2$. This assumes that the pitch moves in the positive direction, meaning that the air drag is in the negative direction. Unfortunately, this equation requires calculus to solve. As seen in the calculus box that follows, the solution to the general equation $a = -kv^2$ gives velocity $v = \dfrac{v_0}{1 + kv_0t}$ and position $x = \dfrac{1}{k}\ln(1 + kv_0t)$. We first need to know how long it takes for the ball to travel 55 feet; we can then plug this time into our velocity equation. To get $55 = \dfrac{1}{k}\ln(1 + kv_0t)$, we need $55k = \ln(1 + kv_0t)$ or $1 + kv_0t = e^{55k}$. Then $kv_0t = e^{55k} - 1$ or $t = (e^{55k} - 1)/(kv_0)$. With $k = 0.00165\ \text{ft}^{-1}$ and $v_0 = 139.3$ ft/s, we get $t \approx 0.413$ s. At this time, the velocity is $\dfrac{v_0}{1 + kv_0t} \approx 127.2$ ft/s or 86.7 mph. Kershaw's fastball loses 8.3 mph, or just under 9% of its speed.

One consequence of this loss in speed has to do with radar guns. Clearly, the point at which the pitch speed is measured can drastically change the reading on the radar gun.

Calculus Box: Solving for Velocity

In Example 1.5, we needed to solve the equation $a = -kv^2$. Given that $a = \dfrac{dv}{dt}$, what we need to solve is the differential equation $\dfrac{dv}{dt} = -kv^2$. This is a *separable* differential equation; its solution technique is covered in many calculus books. (See Smith and Minton.) As the name implies, we want to separate the variables v and t, so we rewrite the equation as $-\frac{1}{v^2}\frac{dv}{dt} = k$ and then integrate.

$$\begin{aligned}\int -\frac{1}{v^2}\frac{dv}{dt}dt &= \int kdt \\ \frac{1}{v} &= kt + c \\ v &= \frac{1}{kt + c}\end{aligned}$$

where c is a constant to be evaluated using the initial condition $v(0) = v_0$. With $t = 0$ and $v = v_0$ our equation becomes $v_0 = \frac{1}{c}$, and we get $c = \frac{1}{v_0}$. Substituting this in, we get

$$v = \frac{1}{kt + 1/v_0} = \frac{v_0}{kv_0 t + 1}$$

as stated in Example 1.5. To get the position function, we use the fact that $v = \frac{dx}{dt}$ and so

$$x = \int vdt = \int \frac{v_0}{kv_0 t + 1}dt = \frac{1}{k}\int \frac{kv_0}{kv_0 t + 1}dt = \frac{1}{k}\ln|kv_0 t + 1| + c$$

where c is again a constant to be evaluated. Since $x(0) = 0$, we have $0 = \ln(1) + c = c$ and so $c = 0$. We are left with $x = \frac{1}{k}\ln(1 + kv_0 t)$ as desired.

Bending with Bubba: Magnus Force

Bubba Watson was in a playoff for the 2012 Masters golf tournament, 163 yards from the hole and aiming about 45 yards left of the green because of trees. His wedge hooked onto the green about 10 feet from the hole, setting up his playoff victory. United States midfielder Jermaine Jones scored an outstanding goal in the 2014 World Cup, bending a ball around a defender into the net. Without the curve on the ball, the shot would have sailed several paces wide of the goal. How did Watson and Jones do it? The role of spin in causing ball trajectories to be curved is examined next.

The spin on a ball in flight is likely to create a force, called the **Magnus force**, that causes the familiar curves of golf balls, soccer/footballs, baseballs, and tennis balls. A basic understanding of the geometry of this force provides insight into the techniques used to control the flights of the balls. Trajectories of balls acted on by gravity, drag, and the Magnus force must be simulated on a computer. We focus on the geometry of the force here.

The Magnus force can be explained by Newton's Third Law: every action has an equal and opposite reaction. Imagine a ball spinning clockwise with air flowing left to right, as in the figure. The motion of the ball pulls the flowing air downward. When the ball pulls the air down, the air will push the ball up. This upward force is the Magnus force. This explanation makes it clear that the greater the spin rate of the ball, the greater the Magnus force.

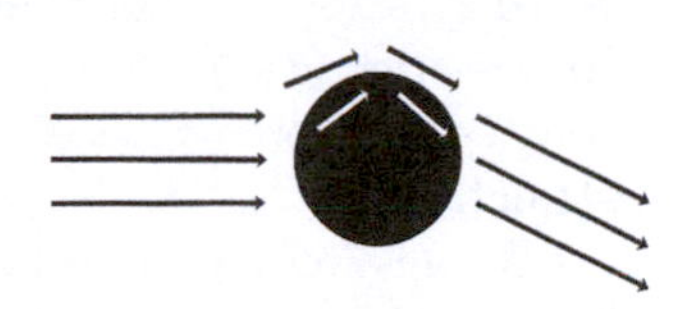

FIGURE 1.3: Air Flow

The Magnus force depends on both the velocity of the ball and its spin vector, defined as follows. Consider a ball with a horizontal velocity directly away from you. As you look at the ball, the back of the ball is rotating from top to bottom as shown in Figure 1.4. This is called **backspin** and is the most common spin in sports. We define a spin vector **s** by first identifying the axis about which the ball rotates. In this example, the spin axis is a left-right horizontal axis through the center of the ball. Now, imagine curling the fingers of your right hand around the front of the ball with your fingers pointing in the direction the ball is spinning. If you extend your thumb, it should point to the right, which is the direction of the spin vector **s**.

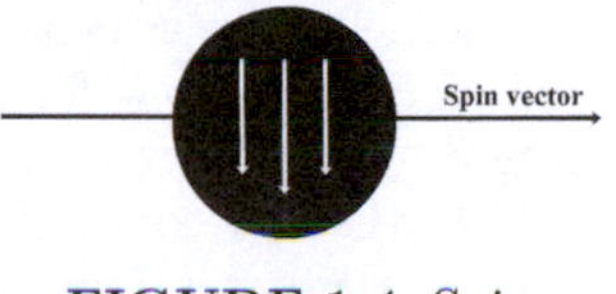

FIGURE 1.4: Spin

The direction of the Magnus force is perpendicular to both the spin vector (to the right) and the velocity vector (pointing away from you). The directions "up" and "down" both qualify; we can determine which one is correct using another right-hand rule. Point the index finger of your right hand in the direction of the spin vector, holding your hand in such a way that you can curl your bottom three fingers toward the velocity vector. Your thumb will point in the direction of the Magnus force; in this case, up. (In calculus terms, if **s** is the spin vector and **v** is the velocity vector, the Magnus force is in the same direction as the cross product $\mathbf{s} \times \mathbf{v}$.)

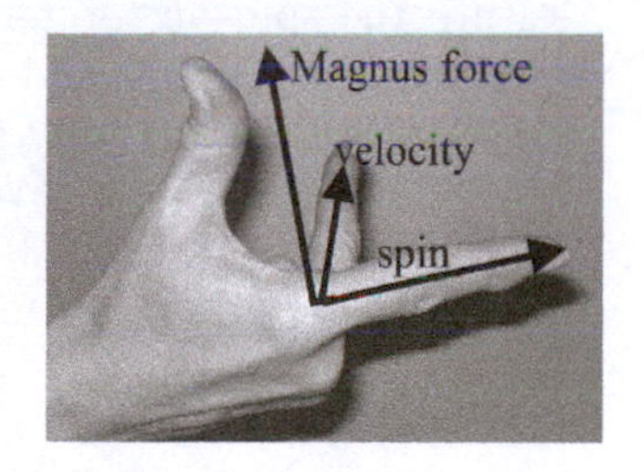

FIGURE 1.5: Right Hand

The first brief take-away is that backspin creates a Magnus force with an upward component. In Figure 1.6, the velocity vector (labeled "Vector") is horizontal and to the right. In this orientation, backspin means that the ball rotates counterclockwise. The Magnus force is straight up.

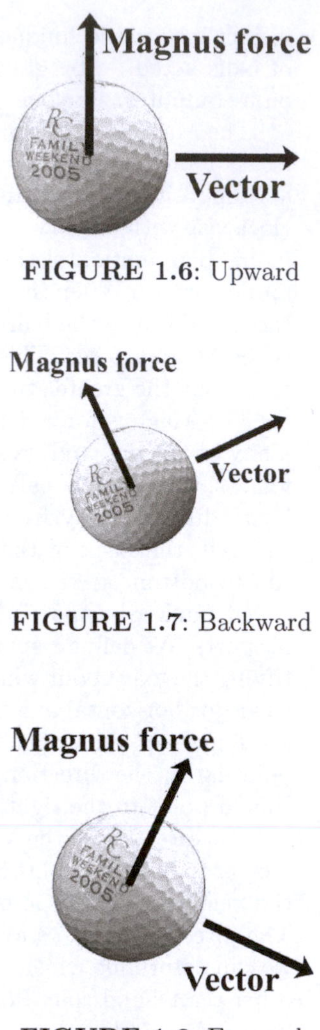

FIGURE 1.6: Upward

FIGURE 1.7: Backward

FIGURE 1.8: Forward

Because of the upward Magnus force, the velocity will not remain horizontal, but will gain a vertical component. The Magnus force and velocity vector remain perpendicular throughout the ball flight, so a change in velocity direction is accompanied by a change in Magnus force direction. Tilt the velocity vector as the ball goes up and the Magnus force tilts backwards as in Figure 1.7.

As the ball comes down, the Magnus force tilts forward as in Figure 1.8. The Magnus force is now pointing up and forward, giving the ball a positive horizontal acceleration. This has implications in the catching of high pop-ups in baseball, as we explore in exercise 1.27.

Figures 1.6-8 can also help us understand what happens if the spin vector has a sideways component. You will need to mentally substitute the spin vector for the velocity vector shown in the Figures, and also assume that the velocity vector is into the page.

So, how do Bubba Watson and Jermaine Jones bend their shots? Pure backspin can be created if Watson's golf club (or Jones's foot) slides directly under the ball. Because the club is coming downward and hitting the ball below center, during the brief time that club and ball are in contact the ball will start to roll up the club; this creates the ball's spin. With pure backspin, we have seen that the Magnus force will be upward (and backward or forward depending on whether the ball is going up or down).

If, however, the clubhead is tilted or the swing path is crooked, the ball will not roll straight up (vertically), but instead will roll some to the side; the ball has some sidespin. Now, assume that the balls in Figures 1.6-8 have velocity into the page and "Vector" shows the direction of the spin vector. Watson swings so that at impact the club slides under the ball and is moving from right to left. The ball rolls up the club and to the right, producing a spin vector like that of Figure 1.8. The resulting Magnus force will bend the ball to the right, as we saw at the Masters. ESPN's *Sport Science* estimates the axis of rotation was tilted by 38 degrees, creating the severe rightward curve that he needed. Jermaine Jones, kicking with his right foot, created a spin vector like that of Figure 1.7, and his shot curved to the left.

Example 1.6 For the following sports situations, identify a possible spin direction, Magnus force direction, and resulting curve of the ball. (a) tennis serve, (b) baseball fastball, (c) baseball curveball, (d) golf outside/in swing.
Solution. Answers will be given for right-handed players. (a) Spin on a tennis serve can come from the motion of the racket and from the ball toss. Imagine a ball dropping from a large height and then contacting a racket squarely. Because the ball is moving downward, as it contacts the strings of the racket it will roll down the racket. This is the exact opposite of the spin in Figure 1.4 (and is called **topspin**). The spin vector will be to the left and the Magnus force downward (for a serve moving horizontally). A right-handed player can easily cause the racket to be moving from left to right as it strikes the ball. This tilts the spin axis as in Figure 1.7 and the ball will slice to the left. (b) A fastball is released by letting the ball roll off of the index and middle fingers. The spin is backspin as in Figure 1.6, with an upward Magnus force. A right-handed pitcher's arm is likely to not move in a purely "overhanded" fashion, but have some right to left movement. This will cause the two fingers to be tilted to the right, creating a spin as in Figure 1.8 (spin vector down and right), and a Magnus force up and right. This creates some movement to the right, as in the "tailing fastball" made famous by Greg Maddux and others. (c) The curveball is released with a snap of the wrist that causes the back of the ball to rotate upward (topspin again) and (for a righthander) to the right. This makes the spin vector point up and to the left, with a Magnus force that is down and to the left. The curveball will move a small amount to the left, but its main movement is downward. (d) A good golf swing always goes underneath the ball, so there is always backspin. For a righthanded golfer, outside/in means that the club is moving somewhat from the right to the left as it moves into the ball. The result is seen in Figure 1.8, with spin vector down and right and Magnus right up and right. This is the classic slice that plagues so many golfers.

Smiling with Dimples

Golf balls have 300-400 dimples, small indentations that cover the ball in a symmetric pattern (most asymmetric patterns are illegal, as an asymmetry may prevent the ball from curving). Historically, golfers had noticed that golf balls flew farther and straighter after getting battered and dented. Golf ball manufacturers stepped in and offered "pre-dented" balls that revolutionized the game. The advantages of dimpled golf balls can be stated in terms of the forces on the ball. A dimpled ball traveling 150 mph with 3000 rpm of backspin has less than half the air drag of a smooth ball; this may seem counterintuitive, but the roughness of the dimpled ball creates a turbulent wake that results in a shorter separation of streamlines and less drag. The television show *Mythbusters* created a dimpled car that got demonstrably better gas mileage.

But, that is not the only advantage of the dimpled ball. At 150 mph and 3000 rpm, the dimpled ball has four times the lift force (upward Magnus force) of a smooth ball. The increased Magnus force lifts the ball higher, giving it time to travel farther, and providing spin to create consistent and controllable trajectories. New golf ball designs go through extensive aerodynamic testing to create ideal drag and lift profiles.

Calculus Box: A General Model of a Ball in Flight

If you mathematically model the forces of gravity, air drag, and Magnus force, you have a highly accurate equation for the flight of a ball. You also have an equation that cannot be solved (meaning that a solution cannot be found in terms of the basic elementary functions that you commonly see in mathematics courses). In our digital age, this is not a deal-killer. We can throw the equations into Mathematica or other software and get a nice graphical "solution" that gives us important information.

The most common form for air drag is $-kv^2$ where v is the speed of the ball and the minus sign indicates that the force is in the direction opposite that of velocity. In vector form, this becomes $-k_d |\mathbf{v}| \mathbf{v}$ where $\mathbf{v}$ is the velocity vector. The Magnus force is generally given as $k_m \mathbf{s} \times \mathbf{v}$ where $\mathbf{s}$ is the spin vector described above. A large complication that is disguised by these expressions is that k_d and k_m are, in general, not constants. They can depend on speed, spin rate, and various environmental factors. The drag coefficient k_d, for example, may depend on speed in a complicated fashion, where there is a sudden decrease in drag at a critical speed. Even treating k_d and k_m as constants, the resulting differential equation

$$\frac{dv}{dt} = -g - k_d |\mathbf{v}| \mathbf{v} + k_m \mathbf{s} \times \mathbf{v}$$

is highly coupled and therefore difficult to solve. Here, "coupled" means that the x-, y-, and z-dimensions do not split off into separate equations. The actual components for a baseball can be found in exercise 1.45. For now, notice that $|\mathbf{v}| = \sqrt{v_x^2 + v_y^2 + v_z^2}$ and the x-component of $\mathbf{s} \times \mathbf{v}$ is $s_y v_z - s_z v_y$ so that you must know v_y and v_z to solve for v_x (and you need v_x and v_z to solve for v_y).

The Effects of Drag and Lift

There are numerous software packages that can give you nice graphs representing solutions of the above equations. Below, we look at graphs of baseballs in flight to get an idea of the importance of drag and lift (Magnus force) in baseball.

Before doing so, we want to get an idea of what the software is doing to create these graphs. Having some idea of what is going on "behind the curtain" helps us to understand the limitations of the software and be more intelligent users of the technology. Plus, it is fun to know what's going on!

This is a fully three-dimensional model of motion. Let's start by orienting ourselves. For baseball, one choice is to have the batter at the origin, the y-axis horizontal through the batter and the pitcher, the x-axis horizontal at a right angle to the y-axis, and the z-axis vertical. We use the general equations from above with appropriate values for the coefficients of $k_d = 0.0018$ and $k_m = 0.000064$ (which we get from Watts and Bahill). To make life a little simpler, let's focus on the equation for acceleration in the z-direction.

$$a_z = -32 - 0.0018 v_z \sqrt{v_x^2 + v_y^2 + v_z^2} + 0.000064(s_x v_y - s_y v_x)$$

where s_x is the x-component of the spin vector in units of rpm.

For initial values of the variables, let's consider a ball hit directly to center field at a height of 3 ft with initial speed of 160 ft/s at an angle of 30 deg above horizontal, and with backspin of 4000 rpm. Then the initial position is $x(0) = 0$ ft, $y(0) = 0$ ft, and $z(0) = 3$ ft. The initial velocity is $v_x(0) = 0$ ft/s (the ball is hit directly toward center field), $v_y(0) = 160 \cos(30°) \approx 138.6$ ft/s, and $v_z(0) = 160 \sin(30°) = 80$ ft/s. The initial values for spin are $s_x = 4000$ rpm, $s_y = 0$ rpm, and $s_z = 0$ rpm (if the spin is pure backspin, the spin vector points directly to the right, the positive x-direction).

Let's substitute these values in and compute the initial acceleration in the z-direction. Note that the square root term represents speed, so instead of trying to compute the term as listed we can simply substitute in the initial speed of 160 ft/s. We have -32 ft/s^2 from gravity, $-.0018(80)(160) \approx 23$ ft/s^2 from drag, and approximately $.000064(4000)(138.6) \approx 35.5$ ft/s^2 from lift.

Notice that at the beginning of the flight of the baseball, lift from the Magnus force is actually larger than the gravitational pull. The drag force is more than 70% of the force due to gravity. In many sports situations, it is not at all reasonable to ignore drag or lift.

Back to the calculation: if we add the pieces together, we get an acceleration in the z-direction of about -19.5 ft/s^2. Because this acceleration is not constant, we cannot use it to compute velocity exactly. However, if the acceleration is approximately constant for a tenth of a second, then the change in velocity for that tenth of a second is approximately acceleration times change in time and

$$v_z(0.1) \approx v_z(0) + 0.1a_z(0)$$

from Table 1.1. Substituting in, we get $v_z(0.1) \approx 80$ ft/s + (0.1 s)(−19.5 ft/s^2) = 78.05 ft/s. Given a z-velocity, we can estimate the z-position from $z(0.1) \approx z(0) + 0.1v_z(0.1)$ and get $z(0.1) \approx 3$ ft + (0.1s)(78.05 ft/s)≈ 37.8 ft.

There are numerous details to which you can reasonably object. Before doing so, let's review the general procedure just outlined. We pick a time step dt (we used $dt = 0.1$ above) and then

1. Start with the current values for velocity and position.
2. Compute the acceleration.
3. Use acceleration to update velocity (as if acceleration were constant).
4. Use velocity to update position (as if velocity were constant).

Now, to address some objections that you may have. The assumptions that acceleration and velocity are nearly constant may not be acceptable (in fact, in our calculation above, velocity changed from 80 ft/s to about 78 ft/s). If we decrease the value of dt, however, the assumption should be more accurate. Theoretically, we can decrease dt until we are happy with the assumption. In practice, the smaller you make dt the more calculations have to be made to reach a target value of t (for example, if the flight time of our baseball is 4 seconds, with $dt = 0.1$ we need 40 updates to reach $t = 4$; with a smaller $dt = 0.01$ we would need 400 updates). In our sample calculation, we used the velocity update of 78.05 ft/s to estimate the height. We could have used the initial velocity of 80 ft/s or (even better) an average of the two. These are issues that you can explore in a course in the field of *numerical analysis*.

The following graph shows the flight of the baseball described above under three assumptions.

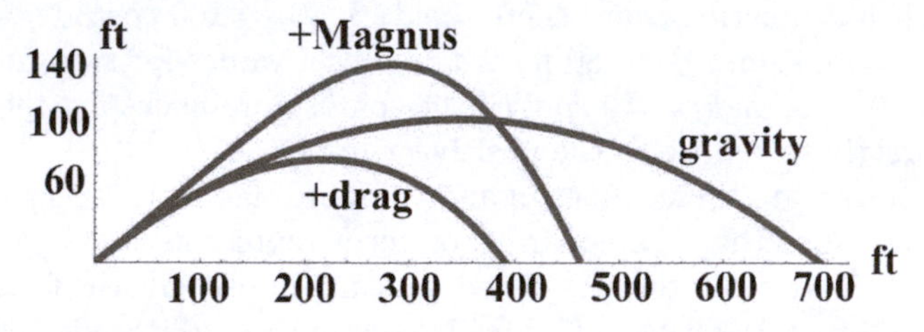

FIGURE 1.9: Effect of Drag and Magnus Forces

The graph labeled "gravity" shows the trajectory of a ball acted upon by gravity only (no drag, no lift). This ball is launched at 160 ft/s (about 109 mph) at an angle of 30 deg. Its range is about 700 ft. How much effect does drag have? The graph labeled "+drag" shows the trajectory of a ball acted upon by gravity and air drag. Its range has been reduced by almost half, down to 385 ft. And what about the Magnus force? The graph labeled "+Magnus" shows the trajectory of a ball acted upon by gravity, air drag, and the Magnus force. The lift force restores some distance, boosting a 385-foot out to a 460-foot home run. The times of flight for the three hits are 5 s, 4.2 s, and 6.8 s, respectively. The backspin on the hit gives the ball extra height, extra hang time, and extra distance.

A final note on the graphs in Figure 1.9 has to do with symmetry. The "gravity" graph has symmetry; the flight up is the same as the flight down. In the "+drag" graph, you may notice that the flight down is shorter than the flight up; since drag is reducing the speed, less distance is covered later in the flight. The "+Magnus" graph is also asymmetric, with the peak of the graph occurring at about the 60% mark of horizontal distance.

Figure 1.10 shows a side view of three baseball pitches, to illustrate the effect of spin. All three graphs show 90-mph pitches. The middle graph includes the effects of gravity and air drag, but no Magnus force. The top graph shows the lift created by 2400 rpm of backspin on a fastball, while the lowest graph illustrates a curve ball with 2400 rpm of topspin.

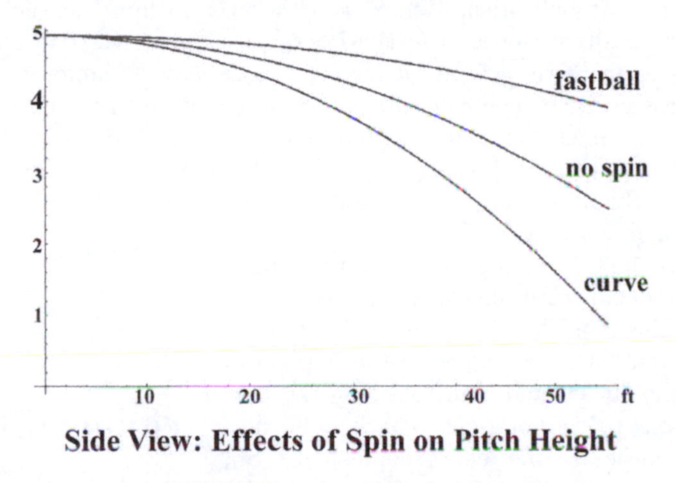

FIGURE 1.10: Effect of Spin

The middle graph is included for reference. As we see in the next section, it does not represent a baseball pitch thrown with no spin.

Knuckling Down

The knuckleball is a baseball pitch known for its contrariness. Its motion fools batters, catchers (Bob Uecker famously said that his technique for catching the knuckleball was to wait for it to stop rolling and then pick it up), and umpires. Even its name is contrary: the ball is not gripped with the knuckles, but with the fingertips (making the knuckles visible to the batter). Knuckleballs are not thrown hard (60-80 mph), have very little spin, and are said to dance around: the web site FanGraphs has several excellent video clips of R.A.

Dickey's knuckleball moving first one way and then the other. The explanation of how the knuckleball works is also unusual.

A baseball has raised stitches holding its cover together. The stitches are raised enough to create a lift force, much like the curved wing of an airplane. Watts and Bahill took wind tunnel measurements of the force for different orientations of the ball. If θ measures the angle of rotation of the ball from some initial orientation, the lateral force on the ball is approximately $-0.1\sin(4\theta)$ lb. (Note that $\theta = 2\pi \approx 6.3$ radians represents one rotation of the ball.)

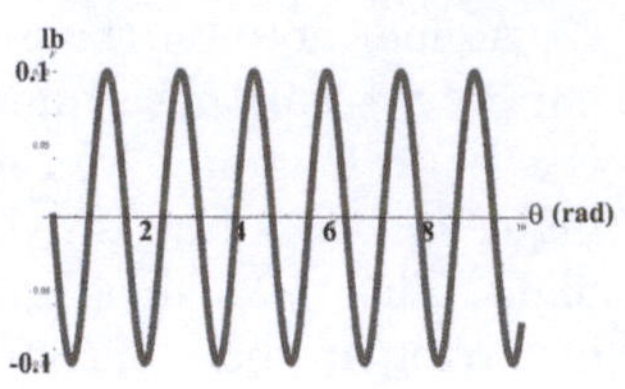

As the ball rotates, the lateral force on the ball due to the stitches oscillates. In this case, lateral force means side-to-side or left-to-right from the batter's perspective. At ball orientation $\theta = 0$ there is no force on the ball. As it rotates to positive values of θ, the lateral force goes negative and the ball will accelerate to the left. As θ increases past $\theta = \frac{\pi}{4}$, however, the lateral force turns positive. The ball now has an acceleration to the right; the ball's leftward movement slows, and the ball may even start moving to the right (as seen at FanGraphs). Following the force in this way can be awkward. As seen in the calculus box below, we can solve for the lateral (left-right) position of the ball and learn more.

Figure 1.11 (*not* to scale) shows the path of a knuckleball with almost no spin ($\omega = 0.1$ rad/s). This is a blimp's-eye view of the pitch from above, with the pitcher on the horizontal axis to the far left and home plate on the horizontal axis to the far right. The ball finishes about 2 inches to the left of home plate. So, this pitch behaves like a 65 mph curveball without much break.

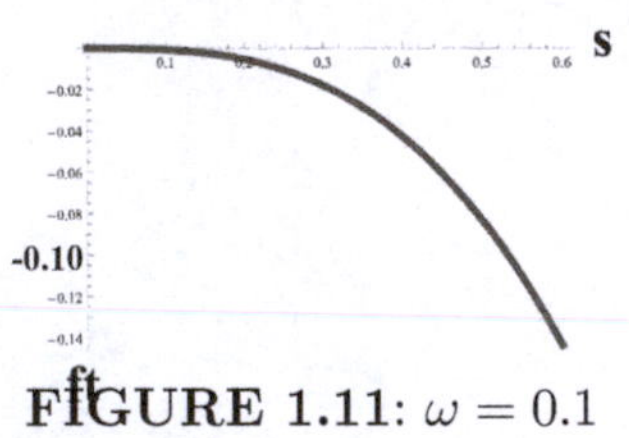

FIGURE 1.11: $\omega = 0.1$

For Figure 1.12, the spin rate is increased to $\omega = 2$ rad/s (about a quarter turn of the ball from pitcher to home plate over 0.65 seconds) and the trajectory changes dramatically. The vertical scale is in feet, so this pitch moved 0.3 feet or 3.6 inches to the left before turning around and coming back to the center of the plate.

For Figure 1.13, the spin rate is further increased to $\omega = 10$ rad/s (about one full rotation of the ball) and a different picture emerges. It is true that the ball is oscillating very rapidly, but the movement is nearly imperceptible and at 65 mph would be unlikely to confuse a major league batter.

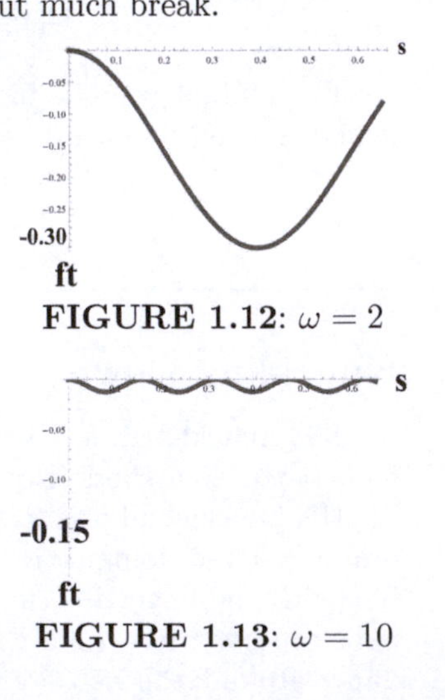

FIGURE 1.12: $\omega = 2$

FIGURE 1.13: $\omega = 10$

We see that the knuckleball is extremely sensitive to the spin rate, with the best results occurring when the ball rotates between one-quarter and one-half of a turn on the way to the plate. This gives some idea of the difficulty of throwing the pitch effectively, and why so few pitchers have had extended success with the knuckleball.

Calculus Box: Lateral Position of a Knuckleball

With spin rate ω rad/s, the lateral position $x(t)$ of a pitch thrown directly toward home plate obeys Newton's Second Law

$$m\, x''(t) = -0.1\ \sin 4\theta$$

Tracking x in feet, we use a ball mass of $m = 0.1$ slug. If the ball spins at a rate of ω rad/s, then $4\theta = 4\omega t + \theta_0$ rad, for some initial orientation θ_0. Our equation becomes

$$x''(t) = -10 \sin(4\omega t + \theta_0)$$

Integrate this with respect to t to find the lateral velocity. Thus

$$x'(t) = \frac{10}{4\omega} \cos(4\omega t + \theta_0) + c$$

where the integration constant c can be determined from the initial condition $x'(0) = 0$ (the ball is thrown directly at home plate). We find $c = -\frac{10}{4\omega} \cos\theta_0$ and so

$$x'(t) = \frac{5}{2\omega} \cos(4\omega t + \theta_0) - \frac{5}{2\omega} \cos\theta_0$$

We integrate one more time with respect to t to get

$$x(t) = \frac{5}{8\omega^2} \sin(4\omega t + \theta_0) - \left[\frac{5}{2\omega} \cos\theta_0\right] t + c$$

where the integration constant c is determined from the initial condition $x(0) = 0$. We find that $c = -\frac{5}{8\omega^2} \sin\theta_0$ and conclude that

$$x(t) = \frac{5}{8\omega^2} \left[\sin(4\omega t + \theta_0) - \sin\theta_0\right] - \left[\frac{5}{2\omega} \cos\theta_0\right] t$$

for a general equation of the knuckleball. Figure 1.10 uses $\theta_0 = 0$ and Figures 1.11 and 1.12 use $\theta_0 = \frac{\pi}{2}$.

Exercises

In exercises 1.1-12, assume that gravity is the only force. Ⓣ refers to thinking problems, conceptual problems requiring no calculations. Ⓒ refers to problems requiring calculus or significant computer calculations. Ⓟ refers to projects; these are ideas for further investigation (hints and resources are at the book's web site).

1.1 An object is dropped from a height of 100 m. Find its velocity and position at times $t = 1$, $t = 2$, and at impact.

1.2 Find the impact speed for divers dropping from heights of (a) 32 ft and (b) 128 ft. (c) Fill in the blank, and prove that you are correct: If height is multiplied by 4, impact speed is multiplied by ___ .

1.3 An object falls from a height of 400 m. Find its position at times $t = 1$, $t = 2$, $t = 3$, and $t = 4$. Compute $dp_1 = p(1) - p(0)$, $dp_2 = p(2) - p(1)$, $dp_3 = p(3) - p(2)$, and $dp_4 = p(4) - p(3)$. Show that $dp_2 = 3dp_1$, $dp_3 = 5dp_1$, and $dp_4 = 7dp_1$. Conjecture values of dp_5 and dp_6. To explain this pattern, compute $2^2 - 1^2$, $3^2 - 2^2$, $4^2 - 3^2$, and so on.

1.4 A diver jumps from a height of 10 m with an initial (upward) velocity of 1 m/s. Ignoring horizontal motion, (a) find the diver's velocity and position at times $t = 1$ and $t = 2$. (b) Find the diver's maximum height, time to impact, and velocity at impact.

1.5 Dwight Howard set a record by making a mark on a backboard 12'6" above the ground. His vertical jump was measured at 39.5 inches. Find (a) the required initial velocity in ft/s and (b) Howard's hang time. (c) The episode "Flight" of the television series *Sport Science* states that a hang time of one second is the limit of human abilities. If this is true, what is a human's maximum vertical jump?

1.6 A football punt has a hang time of 5 seconds. Ignoring horizontal motion, find (a) the required initial velocity in ft/s and (b) the maximum height of the football.

1.7 In Example 1.3, a 25-foot shot is launched with initial speed 30 ft/s at an angle of 50 deg. (a) Is the shot good if the angle is 45 deg? (b) Is the shot good if the speed is 31 ft/s (angle of 50 deg)?

1.8 In Example 1.3, use trial and error to approximate to the nearest one-tenth s_1 and s_2, the smallest and largest speeds, respectively, for which the shot is good.

1.9 Show that a 25-foot shot launched with initial speed 30 ft/s at an angle of 48.5 deg is good. Show that the shot reaches basket height at a shorter distance for both angles of 48 and 49. Conclude that at speed 30 ft/s it is not possible to hit the back rim. This is a characteristic of the launch speed that has the largest margin of error in angle: it is the smallest launch speed to reach the center of the basket. Explain why this could be called the "best" launch speed, and explain how a shooter could "feel" this ideal speed.

1.10 A shortstop throws to first base from 80 ft away, releasing the ball horizontally from a height of 5 ft with speed 80 mph. (a) What happens to the throw? (b) By trial and error, find the angle such that the ball reaches first base at height 5 ft. At this angle, how far above first base is the ball aimed?

1.11 A tennis serve launches a ball with initial speed 120 mph from a height of 10 feet at an angle of 7 deg below horizontal. To be good, the serve must clear a 3-foot-high net that is 39 ft away, and must hit the ground on or before the service line 60 ft away. (a) Is this serve good? (b) By trial and error, find the angles such that the ball clips the top of the net and the ball lands on the service line; the difference in angles is the margin of error at 120 mph. (c) Find the margin of error of a 135 mph tennis serve. Does the extra speed increase or decrease the margin of error?

1.12 A baseball is hit from a height of 3 ft with initial speed 130 ft/s at an angle of 25 deg above the horizontal. Will the ball clear a 6-ft wall located 400 ft away?

1.13 A free kick is taken from 25 yd out from a spot even with the right post. The initial speed is 90 ft/s and the velocity is 4 deg to the right of the post. The kicker puts sidespin on the ball that results in a constant acceleration of 16 ft/s^2 to the left. (a) If the goal is 24 feet wide, will this shot be on goal? (b) If there is a wall 10 yd from the kicker whose rightmost defender extends one foot to the right of the goal, does the kick hit the wall?

1.14 Suppose that a ball is designed with a 23% reduction in constant k from the ball in Example 1.4. By how much is the terminal velocity changed?

1.15 The terminal velocity of a tennis ball is 70 mph. Find the constant k for its drag, and estimate the speed of a 135 mph serve after it has traveled the 60 ft to the service line.

1.16 The terminal velocity for a badminton shuttlecock is 6.8 m/s. Find the constant k for its drag, and estimate the speed of a (world record) 493 km/hr smash after it travels 10 m.

1.17 For the following sports situations, identify a possible spin direction, Magnus force direction, and resulting curve of the ball. (a) tennis topspin shot, (b) baseball sidearm fastball, (c) baseball fly ball to left field by a lefthanded batter, (d) golf inside/out swing, (e) basketball free throw.

1.18 Here are the 10 m splits for Usain Bolt's world record sprint of 9.58 s in 2009: $(10, 1.89)$, $(20, 2.88)$, $(30, 3.78)$, $(40, 4.64)$, $(50, 5.47)$, $(60, 6.29)$, $(70, 7.10)$, $(80, 7.92)$, $(90, 8.75)$, $(100, 9.58)$. That is, he reached the 10 m mark in 1.89 seconds, the 20 m mark in 2.88 seconds, and so on. Use these to compute his average velocities in each 10 m interval. Estimate his speed at 10 m, 20 m, and so on. Which of these estimates is the least accurate? Why? Estimate his acceleration in each 10 m interval.

1.19 At the 2014 Olympics, the winning time for the men's 1500 m speed skating race was 1:45.01. The winning time for the men's 5000 m speed skating race was 6:10.76. Compute average speeds for each event. Explain why the speed for the 5000 m race is slower. Use these two measurements to predict the average speed for a 3200 m race. The team pursuit is this length; the winning time was 3:37.71. Compute the average speed for this race, and explain why the speed is faster than your prediction. At the 2012 Olympics, winning times for the men's 1500 m and 5000 m runs were 3:34.08 and 13:41.66, respectively. Compute average speeds for these races. Explain in terms of forces why humans can skate faster than they can run. The winning time in the men's 4x400 m relay was 2:56.72. Why is this faster than the winning 1500 m time?

1.20 (a) At the 2012 Olympics, the winning time for the men's 100 m dash was 9.63 s. The winning time for the men's 4x100 m relay was 36.84 s. Compute average speeds for each event. Explain why the speed for the relay race is faster. (b) By contrast, winning times for the 400 m and 4x400 m races were 43.94 s and 2 min, 56 s, respectively. Compute average speeds, and explain why the relay is slower.

1.21 At the 2012 Olympics, the winning time for the men's 1500 m race was 3:34. The winning time for the men's 4x400 m relay was 2:56. Compute average speeds for each event. Explain why the speed for the relay race is so much faster.

1.22 Find world record times for different distances for men's and women's running races, and compute average speeds for each. Try to find a formula for a function f such that $f(d)$ is close to the average speed for a race of length d m.

1.23 A driver averages 140 mph for the first 10 miles of a race and 180 mph for the second 10 miles of the race. What is the average speed for the first 20 miles of the race?

1.24 The drag force is often written as $\frac{1}{2}Ac\rho v^2$ where A is the projected surface area of the object, c is a coefficient depending on numerous factors, ρ is the density of the air, and v is the speed of the object. Tests of footballs have shown that a punt moving in a spiral has a value of Ac that is about 18% of that of an end-over-end kick. Given that air density in Denver is 18% less than in New York, compare the drag on a spiral in Denver to an end-over-end kick in New York.

1.25 Convert ω rad/s to rpm. If a pitch has 2000 rpm of spin and takes 0.45 s to reach home plate, how many rotations does it complete?

1.26 A knuckleball is thrown with the ball gripped by the fingertips. Given that a good knuckleball has little spin, why is this grip good? For pitches for which more spin is better, explain how the ball can be gripped to increase spin.

1.27 Ⓣ Figure 1.1 shows a graph of height h as a function of time t. If gravity is the only force and the ball has a horizontal motion, explain why the graph of h versus horizontal distance x would look the same.

1.28 Ⓣ Look up world records for track races for men and women, and record the year in which the world record was set. In cases where the record was set more than, say, 5 years ago, speculate on why the best athletes in the world have not broken the record recently.

1.29 Ⓣ For a ball launched vertically with backspin, (a) describe the direction of the Magnus force on the way up, at the top, and on the way down; (b) explain why the ball cannot go up (or come down) in a purely vertical direction; (c) explain why baseball infielders on high pop-ups often have to backpedal rapidly to make the catch.

1.30 Ⓣ Filip Bondy's book *The Pine Tar Game* details a baseball game between Kansas City and New York on July 24, 1983. George Brett, after hitting what he thought was a go-ahead home run for Kansas City, was ruled out because the sticky pine tar on his bat had spread too far up the handle. Suppose that the ball hit the bat on one of the areas with pine tar. Explain how the ball could get extra spin and travel farther because of the pine tar.

1.31 (T) In Figure 1.4, the Magnus force direction turned out to be exactly opposite the direction of the motion of the back of the ball (the arrows). This is true in general. Illustrate this for two other arrow directions.

1.32 (T) A soccer player kicks the ball with the instep. For a rightfooted kick, explain in terms of ball spin why the ball typically bends from right to left. If the ball is struck with the outside of the foot, how does the spin change?

1.33 (T) Suppose a golfer takes swings that produce straight shots. If the righthanded player's left foot is moved closer to the ball ("closing" the stance), explain why a hook (right-to-left movement of the ball) is likely to result.

1.34 (T) In 2014, the NCAA changed its baseball specifications to reduce the size of the stitches (from being raised 1 mm to being raised 0.5 mm). The goal was to produce more home runs. Explain in terms of drag force why this might work.

1.35 (T) (a) The spin on a major league pitch can approach 2500 rpm. Explain why the lift force from the stitches is not important. (b) Pitching in Denver (high altitude), explain why you would expect the ball to reach home plate faster but with less movement than at low altitude.

1.36 (T) A sprinter such as Usain Bolt quickly reaches a peak speed. If he maintains that speed, the forces on him must balance. The push of his feet against the track provides a positive acceleration. Give two examples of negative forces that cancel this out.

1.37 (T) In various types of racing, "drafting" is important. Explain in terms of forces what advantage(s) drafting provides.

1.38 (T) In tennis, so-called "spaghetti" stringing styles (now banned) caused the contact time between racket and ball to increase. Discuss the resulting changes to spin rate, the Magnus force, and the playing of the game.

1.39 (T) If a baseball has a terminal velocity of 95 mph, explain why a pitch thrown horizontally at 95 mph loses speed.

1.40 (T) *The Puzzler's Dilemma* presents evidence that swimmers can swim as fast in a thick, high-viscosity medium (e.g., syrup) as they swim in water. Clearly, the drag in a thicker medium is higher than in a thinner medium. Explain what force could compensate for the drag increase and allow swimmers to reach the same velocity.

1.41 (T) Hold a (low-mass) racquetball directly on top of a (high-mass) soccer ball, then drop the two balls simultaneously. The racquetball should bounce high into the air. Use Newton's Third Law to explain why the racquetball bounces so much higher than the soccer ball.

1.42 (T) Imagine a golf putt that is uphill with no break. Suppose the putt starts out slightly off line. Argue that gravity will pull the putt farther off line. Contrast that to the effect of gravity on a downhill putt.

1.43 (C) For a constant acceleration $a = c$, derive the formulas for velocity and position given in Table 1.1. Derive formulas for velocity and position for an acceleration of $a = ce^{-t}$.

1.44 (C) To find the angle of entry in Example 1.3, note that the slope of a curve is given by $\dfrac{dy}{dx} = \dfrac{y'(t)}{x'(t)}$. Find this slope at the entry time $t = 1.29$ and then use the fact that the absolute value of slope equals $\tan\theta$ to find the entry angle θ. To show that the ball does not touch the rim, we need to find the minimum distance from the trajectory to the rim. The distance between a point (x, y) and the front of the rim at $(24.25, 0)$ is $d = \sqrt{(x - 24.25)^2 + y^2}$. We assume that the ball follows the tangent line at the point of entry, which is $y = -0.949(x - 24.9)$. In the expression for d, replace y with $-0.949(x - 24.9)$. Then solve the equation $d'(x) = 0$ to find the x-value of the closest point. Finally, show that the minimum distance is greater than 4.75/12, the margin of error for the shot.

1.45 Ⓒ The function $f(\omega) = \frac{1.7}{\omega} - \frac{5}{8\omega^2}\sin(2.72\omega)$ gives the position in feet to the left or right of the center of home plate of a knuckleball when it reaches home plate. Graph this function, and comment on the spin rate ω that produces the best pitch. Find the value of ω that produces the maximum deviation. Find the limit of the function as ω goes to 0.

1.46 Ⓒ For a ball launched from the ground with initial speed v_0, show that the (gravity-only) maximum range is obtained with a launch angle of 45 deg. If the launch height and landing height are not the same, define α as the angle in degrees above or below horizontal (e.g., the slope of the ground). Show that the maximum range is obtained with a launch angle of $45 + \frac{1}{2}\alpha$.

1.47 Ⓒ If drag is proportional to v instead of v^2, rework Examples 1.4 and 1.5. How much difference is there?

1.48 Ⓒ Simulate baseball hits using equations $v_x'(t) = -0.0018v_x\sqrt{v_x^2 + v_y^2 + v_z^2} + 0.000064(s_y v_z - s_z v_y)$, $v_y'(t) = -0.0018v_y\sqrt{v_x^2 + v_y^2 + v_z^2} + 0.000064(s_z v_x - s_x v_z)$ and $v_z'(t) = -32 - 0.0018v_z\sqrt{v_x^2 + v_y^2 + v_z^2} + 0.000064(s_x v_y - s_y v_x)$.

1.49 Ⓒ Pitchers release the ball at different locations. The average release point is about 55 feet from home plate. Assume that a pitcher releases the ball 53 feet from home with an initial velocity of 93 mph. Compare the time it takes this pitch to reach home to a 95 mph pitch released from the typical 55-foot mark.

1.50 Ⓟ Find equations for tennis serves on the book's web site. Explore the effects of such variables as height of player; location of server along the baseline; amount that server jumps into the court; serving up the middle versus wide; amount of topspin on the ball. Howard Brody has published two excellent books on the science of tennis.

1.51 Ⓟ Find equations for golf shots on the book's web site. Explore the effects of such variables as amount of sidespin; changes in elevation; loft of the club. *Golf By the Numbers* has more on these topics.

1.52 Ⓟ For a track event such as the marathon, look at the changes in times (world record times, or Olympics winning times, or other) over the years and develop a formula to predict times in the future. You may need to take into account changes in equipment and illegal doping practices. Try to determine the limits of human performance. John Brenkus's *Perfection Point* does this type of investigation.

1.53 Ⓟ Experiments to determine drag or lift forces are generally very difficult to complete. However, here is a simple one you can complete with a simple rangefinder that can compute speed. Take a coffee filter and let it drop. It should quickly reach a terminal velocity that you can record. Then repeat the process with a second coffee filter nested inside the first. All elements of the forces on the falling object have been maintained except that mass has doubled. If the drag force is proportional to speed, then the terminal velocity should double. If the drag force is proportional to the square of the speed, then the terminal velocity will increase by a factor of $\sqrt{2} \approx 1.4$.

Further Reading

Peter Brancazio's *Sport Science* is an excellent introduction to the physics of sports.

Robert Watts and Terry Bahill's *Keep Your Eye on the Ball* is the best of several physics of baseball books.

Equations of motion for baseball pitches were adapted from Alan Nathan's and Michael Richmond's excellent web sites.

A search for "Physics of (favorite sport)" will bring up many results. I can recommend *The Physics of Basketball* (Fontanella), *The Physics of Hockey* (Hache), *Football Physics* (Gay), *Newton's Football* (St. John and Ramirez), *The Science of Soccer* (Wesson), *The Physics and Technology of Tennis* (Brody, Cross, and Lindsey), *Golf Science* (Smith), *The Science of Golf* (Wesson), *Golf By the Numbers* (Minton), and *The Physics of NASCAR* (Leslie-Pelecky).

The John Brenkus television series *Sport Science* explored basketball players' phenomenal jumping abilities in episodes called "Flight" and "Skywalking." The Bubba Watson Masters tournament shot and Dwight Howard vertical jump are featured in separate three-minute segments for ESPN.

Data about drag and lift forces for dimpled versus smooth balls can be found in *Golf By the Numbers* and the Titleist web site.

Information about speed skating can be found in *Gliding For Gold* by Denny.

An analysis of times for Usain Bolt's races can be found at http://sportsscientists.com/2009/08/analysis-of-bolts-9-58-wr accessed 7-21-2015.

An article on the NCAA baseball stitches is at http://www.baseballamerica.com /college/ncaa-to-switch-to-flat-seamed-balls-in-2015 accessed 7-21-2015.

The coffee filter experiment is from *The Dick and Rae Physics Demo Notebook.*

Brody's other tennis science book is *Tennis Science for Tennis Players.*

Further suggestions can be found in the notes at the *Sports Math* web site.

Chapter 2

Rotational Motion

Introduction

Roger Federer's serve is poetry in motion, a graceful knee bend and arch of the back followed by a powerful overhead smash of the ball. His elegance belies the difficulty of the traditional service motion. Both arms raise into the air, one tossing the ball and the other cocking the racket behind the head. As the racket unwinds behind the head, the back foot is brought near the service line. Then both feet push the body up and into the court as the racket meets the ball at the top of the swing. Beginners sometimes take shortcuts, such as simply raising the racket up and swinging it forward without cocking it behind the head. Given that all professional tennis players use (roughly) the same technique, there must be advantages to the awkward manipulations of the serve. We explore the physics behind the tennis serve as we look at rotational motion.

Along the way, we will also answer the following questions. Why are backswings in golf, tennis, and baseball pitching important? Why are tennis rackets and golf clubs so large? Do tall golfers have an advantage? What is the optimal length of a baseball bat? How do ice skaters perform fast spins? Why do acrobats pull their bodies into those awkward-looking tuck positions?

Going in Circles

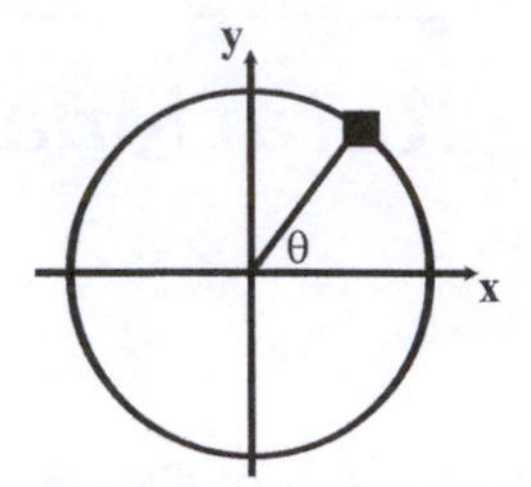

FIGURE 2.1: Angle θ

An object that is spinning in place has rotational motion, but none of the translational movement discussed in Chapter 1. A spinning ball in flight and a basketball player spinning to the basket have both types of motion. To analyze the rotational motion, we track its position using an angle θ (the Greek letter theta). Suppose an object is traveling in a circle. Place imaginary x- and y-axes with the origin at the center of the circle. The line segment from the origin to the object makes an angle θ with the positive x-axis. By tradition, we measure θ in radians, with counterclockwise being the positive direction.

Example 2.1 An object rotates counterclockwise at a constant speed, crossing the positive x-axis at time $t = 0$ and completing one lap every 20 seconds. Find an equation for its angle $\theta(t)$ for any time t.
Solution. With t in seconds, we know that $\theta(0) = 0$, $\theta(20) = 2\pi$ (one lap is 2π radians), $\theta(40) = 4\pi$, and so on. Assuming that $\theta(t)$ is a linear function, we need a line through $(0, 0)$ and $(20, 2\pi)$. The slope is $\frac{2\pi - 0}{20 - 0} = \frac{\pi}{10}$ and so $\theta(t) = \frac{\pi}{10}t$.

In Example 2.1, the slope corresponds to rotational speed, more properly called **angular velocity** (denoted by ω, the Greek letter omega). Then angular velocity is the rate of change of the angular position θ. Completing the trilogy, we define **angular acceleration** α (the Greek letter alpha) as the rate of change of angular velocity. Table 1.1 for translational motion has its rotational equivalent in Table 2.1.

TABLE 2.1: Formulas for Constant Acceleration

angular acceleration	$\alpha = c$
angular velocity	$\omega = ct + \omega_0$
angular position	$\theta = \frac{1}{2}ct^2 + \omega_0 t + \theta_0$

In Example 2.1, the object has constant angular velocity, its angular acceleration is zero, and Table 2.1 shows that θ is a linear function.

There is a simple connection between translational variables and rotational variables. If an object travels in a (two-dimensional) circle centered at the origin with radius r, then the relationships in Table 2.2 hold.

TABLE 2.2: v to ω

$$x = rcos\theta$$
$$y = rsin\theta$$
$$|v| = r|\omega|$$

That is, the translational speed of the object is the rotational speed times the radius of motion. We can use this fact and Table 2.1 to draw a simple inference about tennis serves, golf swings, baseball pitches, and the like.

Example 2.2 Three tennis players start swings with zero angular velocity, applying a constant angular acceleration c rad/s^2. Players A and C use rackets of length L ft while player B's racket is 10% longer. Players A and B rotate through an angle of π radians while player C rotates through an angle that is 10% longer. Compare the final speeds of the ends of the rackets of the three players.

Solution. For player A, $\theta = \frac{1}{2}ct^2$ rad and $\omega = ct$ rad/s. To rotate the racket π radians, it takes time t such that $\pi = \frac{1}{2}ct^2$ or $t = \sqrt{\frac{2\pi}{c}}$ s. At this time, angular velocity is $ct = \sqrt{2\pi c}$ rad/s, and the racket speed is $L\sqrt{2\pi c}$ ft/s. The same calculations hold for player B except that L is replaced by $1.1L$ (the length L plus 10% of L). The racket speed is $1.1L\sqrt{2\pi c}$ ft/s, which is 10% larger than that of player A. For player C, the final angle is not π but 1.1π. This occurs at time $t = \sqrt{\frac{2.2\pi}{c}}$ s, and results in a racket speed of $L\sqrt{2.2\pi c}$ ft/s. Since $\sqrt{2.2} \approx 1.0488\sqrt{2}$, this speed is about 5% higher than that of player A.

Example 2.2 shows that, all things being equal, (1) the longer the racket (or golf club, or baseball bat) the higher the racket speed, and (2) the longer the arc of the swing, the higher the speed. In this case, *all things being equal* means that the angular acceleration remains constant. Unfortunately, sports swings do not have constant acceleration throughout the entire arc of the swing. More importantly, a longer racket is likely to be harder to swing, which probably reduces angular acceleration (see Example 2.4 below). Given these caveats, what can we learn about sports techniques?

The rotation of the tennis racket behind the server's head lengthens the arc of the service motion, thereby allowing for more speed to be generated. The ball is contacted with the server's arm and racket fully extended to maximize the length L; the angular acceleration is aided by the server jumping up a small amount. The tennis serve is designed to maximize ball speed.

Golfers competing in Long Drive championships use preposterously long clubs, the drivers reaching 48" long. The extra length provides extra distance, as in Example 2.2; this comes at the expense of control, but long drive champions must take that risk. Further, long drive champions usually take very big swings (well "past parallel" in golf jargon, with arms extended upward) to increase the length of their swing arc.

Torquing Off Newton

Example 2.2 shows some of the effects of angular acceleration, but how is angular acceleration produced? In Chapter 1, we saw that force is directly related to translational acceleration. We multiply by the radius of the rotational motion to translate rotational variables into translational variables, so it should not be a surprise that we want to look at radius times force. We call this quantity **torque.** In simple terms, torque τ (the Greek letter tau) equals force times distance. Think of a wrench: you apply a force of F lb at the end of a wrench of length L ft, producing a torque of $\tau = FL$ ft-lb that causes a bolt to rotate.

Determining the distance L can be a little tricky. If you push on a door, the torque depends on where and in which direction you push. Pushing on the end of the door directly toward the hinges does not produce a torque (the door won't move); the relevant distance is not measured from hand to hinge, but is the distance from the line of force to the hinge. If you are pushing toward the hinge, the line of force goes through the hinge and the distance is zero. No torque is produced, and therefore no rotation. You know by experience to push on the end of the door at a right angle to the door; this maximizes the distance of the line of force to the hinge, produces the maximum torque, and closes the door most efficiently. (The vector definition $\tau = \mathbf{r} \times \mathbf{F}$, where $\mathbf{F}$ is the force and $\mathbf{r}$ is the vector from the pivot point to the point of application of the force, captures all of the details succinctly.)

Newton's Second Law has the rotational equivalent

$$\tau = I\alpha$$

relating angular acceleration to torque and I, the **moment of inertia (MOI)**. MOI is discussed in detail in the next section; first, let's see a quick example of torque in action.

Example 2.3 A football defensive player grabs a running back by the ankles, applying a weak force of 80 lb for 0.1 s at a distance of 3 ft from the runner's center of mass. If the runner has a (head-over-heels) moment of inertia of 20 ft-lb-s^2, find the runner's angular velocity.
Solution. We compute the torque using the formula $\tau = FL = (80 \text{ lb})(3 \text{ ft}) =$ 240 ft-lb. We then have $\tau = I\alpha$ so that 240 ft-lb $= (20 \text{ ft-lb-s}^2)(\alpha)$ or $\alpha = 12$ rad/s^2. A constant angular acceleration of 12 rad/s^2 for 0.1 s results in an angular velocity of 1.2 rad/s. This is not a very fast rotation rate, but it does mean that the running back's body is rotating about his center of mass. He will fall unless he can apply some sort of balancing force. Even a small force can trip up a runner if applied far from the center of mass.

All About MOI

An important realization is that the question "what is an object's MOI" is meaningless. In Example 2.3, the MOI was specified as a "head-over-heels MOI" because a running back being tripped would tend to rotate head over heels (or is that heels over head?). Just as mass can be defined as resistance to translational acceleration, the moment of inertia can be defined as resistance to angular acceleration. And there's the problem: if the axis of rotation changes, an object's resistance to rotation is likely to change. It is much easier to spin a person about a vertical axis than it is to rotate about the horizontal head-over-heels axis. Thus, a person's MOI about a vertical axis is smaller than the person's MOI about a horizontal axis. This is an important consideration when thinking through rotations in gymnastics, diving, or skating. The lesson is as follows. **An object has a specific mass, but its MOI depends on the orientation of the object and the axis of rotation.**

A further distinction between mass and MOI can be illustrated by the following experiment. Find a sledgehammer; it has a well defined (and large!) mass. Grab it at the end of its handle and try to rotate it end over end (i.e., swing the sledgehammer); the resistance you feel is its large MOI for this rotation. Now, grab the sledgehammer on its metallic hammer end and try to rotate it end-over-end; this is a very easy rotation (small MOI). How does the sledgehammer have different MOIs for two rotations that are both end-over-end? In the first case, the heavy hammer end is a large distance away from the rotation pivot; in the second case, the heavy hammer end is a very small distance from the rotation pivot. So, MOI has to do with mass and the distance that the mass is from the rotation pivot. If the sledgehammer is idealized to a weightless handle and hammer mass concentrated at one point, then MOI equals md^2, where m is the mass of the hammer, and d is the distance between the hammer and the pivot. For sports, the following principle is helpful to keep in mind.

The farther the mass is from the center of rotation, the larger the MOI is for that rotation.

Example 2.4 Compare the MOIs and angular accelerations for the following: (a) a mass of 2 kg rotating 3 m from the center of rotation; (b) a mass of 2 kg rotating 4 m from the center of rotation.
Solution. In this simple case, we can use the formula $I = md^2$ to get (a) $I = 18$ kg-m^2 and (b) $I = 32$ kg-m^2. Moving the mass one meter (33%) farther from the center increased the MOI by nearly 78%! The product of MOI and angular acceleration is torque, so for equal torques the angular acceleration in case (b) is $\frac{18}{32}$ (about 56%) of the angular acceleration in case (a).

Size Is Important

Over the past thirty years, equipment in such sports as golf and tennis has been completely transformed. In particular, golf clubs and tennis rackets are much larger than in the past. In what way is bigger better?

Picking up the story in 1970, a retired aircraft engineer named Howard Head developed tennis elbow. When he hit a tennis ball off-center, his racket would twist and put pressure on the elbow. His engineering solution was to reduce the twisting of the racket by increasing its MOI for off-center hits. The calculation in Example 2.4 shows how to do it: move the mass farther from the center. In this case, most of the mass of the racket head is in the frame, so moving the frame away from the center by enlarging the racket head will increase MOI. As an unexpected consequence, the wooden rackets of the day could not be made with larger heads without becoming too heavy or cracking, so Head designed a large, lightweight metal frame that launched the Prince tennis racket revolution.

Example 2.5 Find the ratio of the MOIs for off-center hits on circular racket frames of inner radii 12 cm and 13 cm, each of which is 1 cm thick.
Solution. The calculation in the calculus box below shows that the MOI is a constant times $f(R) = R^3 + \frac{3}{2}R^2 + R + \frac{1}{4}$. Substitute in $R_1 = 12$ to get $I_1 = f(12)$, proportional to the MOI for the smaller racket. Substitute in $R_2 = 13$ to get $I_2 = f(13)$ for the larger racket. The ratio of the MOIs is given by $\frac{I_2}{I_1} \approx 1.259$. Therefore, enlarging a 12 cm radius frame by 1 cm (8.5%) to a 13 cm radius frame increases its MOI for off-center hits by 26%. Tennis rackets are not circular, but a similarly large increase in MOI for tennis rackets was obtained in the bigger rackets.

Calculus Box: Calculating MOI

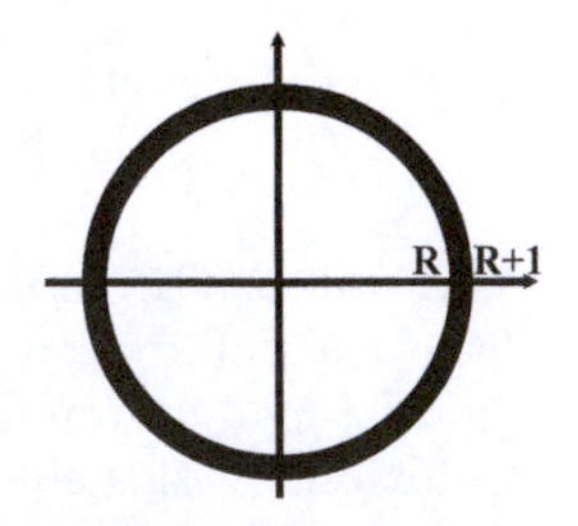

We first approximate a tennis racket frame with concentric circles of radii R and $R+1$ (assuming that the frame is 1 cm thick). We also assume that the frame has a constant mass density c kg/cm^2. The rotation we are concerned with is a rotation about the x-axis, as would occur if the racket were held horizontally and the ball hit above or below the x-axis. The moment of inertia about the x-axis is given by the double integral $\iint_F cy^2 dA$ where F is the frame of the racket.

This integral is best done in polar coordinates, where $y = r\sin\theta$. Then

$$I = \int_0^{2\pi} \int_R^{R+1} c(r \sin\theta)^2 r\, dr\, d\theta = c \int_0^{2\pi} \sin^2\theta\, d\theta \int_R^{R+1} r^3 dr$$
$$= c\pi \frac{1}{4}\left[(R+1)^4 - R^4\right] = c\pi(R^3 + \tfrac{3}{2}R^2 + R + \tfrac{1}{4})$$

for a circular racket frame. More realistic shapes produce more difficult integrals.

Equipment Design

In the early days of metal frames on tennis rackets, players would tape extra weight to the frame. This weight, at a maximum distance from the center of rotation of the racket, further increased the MOI for off-center hits (and may have helped make the racket swing more like a wooden racket). Rackets featuring "perimeter weighting systems" (PWS) began to appear, building extra weight into portions of the frame to improve MOI and swing characteristics.

Similar concerns have impacted golf equipment design. Here, the concern with twisting on off-center hits has less to do with pressure on the golfer's body and more to do with the consequences for the golf ball. If the golf club twists, the initial direction of the ball will change, its spin will change, and power will be lost. Golf equipment advertisements often throw around the initials MOI. Which rotations are they trying to minimize?

Reading left-to-right, the picture to the right shows a driver and two 3-woods. The 3-woods were both good clubs for their days, which were 1995 (far right) and 2010 (middle). The 3-wood grew a lot in 15 years! The 3-wood in the middle is about the size of a 1995 driver, which is dwarfed by the modern driver to its left. Why are larger clubs better? A ball hit "on the toe" by one of these clubs contacts the club outside of the center line perpendicular to the club face. The club will rotate clockwise in response. A larger MOI for this rotation reduces the amount of rotation, and therefore produces a better outcome. The larger club has its mass at a larger distance from the center of rotation, and therefore has the larger MOI for this rotation. Golf clubs, like tennis rackets, grew larger to increase MOI and improve performance. Drivers are now legislated to have a volume of 460 cc or less.

Putters also changed over time. Examine the two putters in the picture to the right. Which one would perform better if the ball were struck above or below center? Look at the backs of the putters, where one is flat and the other curves upward like a parabola. The putter to the right has extra weight on its perimeter at a large distance from the center, giving it a larger MOI for clockwise or counterclockwise rotations caused by mishits.

The benefit to the athlete is in forgiveness. An imperfect swing that causes a tennis or golf ball to be hit off-center will produce less rotation of the club, with the resultant shot looking more like it would have if the ball had been hit perfectly in the center of the club. One way of summarizing the previous sentence is to say that the new rackets/clubs have "larger sweet spots." This manages to convey useful information without really making any sense. If "sweet spot" refers to the best point on the racket/club to hit the ball, then the sweet spot can move (a focus of modern golf club design) but you can't really make a point any larger. Nevertheless, the implication of the advertising campaign is clear enough. The new rackets/clubs, by virtue of large MOIs and other characteristics, have larger regions that will produce "good" shots that are not much worse than your best. Since most of us spend most of our time taking imperfect swings, this is good news.

Supercats and Tamedogs

Snowboarders, gymnasts, and divers all do acrobatics that involve rotational motion. While the terminology of snowboarding can be more entertaining than enlightening (a supercat is a double backflip done on a straight jump, and a tamedog is a frontflip performed on a straight jump), acrobatics involve two main types of rotation. A rotation of the body about the vertical axis (that is, rotating about an axis through the head and down between the feet) is a "twist" and a rotation about a horizontal axis (that is, rotating head over heels) is a "somersault" or "flip." Somersaults, in turn, can be done in different positions. The "tuck" has the body balled up with hands tightly grabbing lower legs or ankles. The "pike" has the body bent in half at the waist, with hands grabbing the back of the knees. The "layout" has the body fully extended.

Can you order the MOIs for the tuck, pike, and layout? In the tuck position, as much of the body as possible is pulled in close to the center of mass (the center of rotation). While this is not an elegant position, it does minimize MOI and facilitate rapid rotations. The layout keeps as much of the body as far from the center of mass as possible, maximizing the MOI and making multiple somersaults difficult. The pike position is in between. In *The Dynamics of Sports*, Griffing gives typical values for a gymnast as 3.5 kg-m^2 for the tuck position, 6.5 kg-m^2 for the pike position, and 15.0 kg-m^2 for the layout position. Dives and gymnastic moves are given a degree of difficulty rating based, in part, on the MOIs of the rotations.

Example 2.6 For a given torque, compare the times needed to complete a triple somersault in the tuck position versus a double somersault in the pike position.

Solution. Let the subscript *tp* refer to the tuck position and the subscript *pp* refer to the pike position. Then since the torques are the same, $\tau = I_{tp}\alpha_{tp} = I_{pp}\alpha_{pp}$ or $3.5\alpha_{tp} = 6.5\alpha_{pp}$ where we use the MOIs from above. We get $\alpha_{pp} = \frac{3.5}{6.5}\alpha_{tp}$ or $\alpha_{pp} \approx 0.54\alpha_{tp}$. Assuming that the torque was applied for the same amount of time t, then angular velocities are given by $\omega_{tp} = \alpha_{tp}t$ and $\omega_{pp} = \alpha_{pp}t \approx 0.54\alpha_{tp}t = 0.54\omega_{tp}$. That is, the angular velocity in the pike position is only slightly more than half the angular velocity in the tuck position, so it takes nearly twice as long to complete a somersault in the pike position. Thus, for a given torque, two somersaults in the pike position take almost as long as four somersaults in the tuck position. Conversely, a triple somersault in the tuck position takes about the same amount of time as $3(0.54) = 1.62$ somersaults in the pike position. The double somersault in the pike position is harder.

Keeping the Momentum

The concept of **angular momentum** gives us an easier way to do the comparison in Example 2.6. Angular momentum L is defined by

$$L = I\omega$$

and the torque equation $\tau = I\alpha$ is more generally interpreted as "torque equals the rate of change of angular momentum" (since α is the rate of change of ω). Thus, $\tau \approx \Delta L/\Delta t$. Since the two divers in Example 2.6 apply the same torque over the same time, their changes in angular momentum are the same, and we can jump directly to $I_{tp}\omega_{tp} = I_{pp}\omega_{pp}$ or $3.5\omega_{tp} = 6.5\omega_{pp}$.

A further consequence of $\tau \approx \Delta L/\Delta t$ is that when there is no torque, angular momentum does not change. This is known as *conservation of angular momentum.*

In the absence of torque, angular momentum remains constant.

We can now follow a diver or gymnast to a safe landing. While the tuck position is used to complete several rotations, the landing is done in the layout position (unless you're doing a cannonball). What is the effect of the diver or gymnast unfolding from the tuck position into the layout position? With the athlete in the air, there is little torque being applied (perhaps a small amount of air drag; let's ignore that). If there is no torque, then angular momentum is conserved. If ω_1 is the angular velocity in the tuck position and ω_2 is the angular velocity in the layout position, then $3.5\omega_1 = 15.0\omega_2$ or $\omega_2 = \frac{3.5}{15}\omega_1 \approx 0.23\omega_1$. Just by changing positions in the air, the athlete's rotation rate has been reduced by 77%, which makes sticking the landing easier.

A flashy application of conservation of angular momentum is performed by figure skaters. A figure skater starts a stationary spin with legs spread apart and arms outstretched. Spinning about a vertical axis, the skater's MOI is relatively large. What happens if the arms are pulled tight to the body and legs brought together? The skater's actions reduce the MOI for the spin, and the effect can be crowd-pleasing.

Example 2.7 A figure skater pulls in her arms and legs to reduce her MOI for the spin by a factor of four. What is the effect?
Solution. By conservation of angular momentum, assuming no torques (so ignoring air drag and ice friction) the product $I\omega$ remains constant. If the subscript b denotes the beginning of the spin and the subscript n denotes the new spin after the change in position, then $I_b\omega_b = I_n\omega_n$. We assume that $I_n = \frac{1}{4}I_b$, so that $I_b\omega_b = \frac{1}{4}I_b\omega_n$ from which we conclude that $\omega_n = 4\omega_b$. A reduction of MOI by a factor of 4 results in an increase in spin rate by the same factor of 4. The skater spins faster. The spin rate can be dramatically manipulated by careful changes in MOI.

Conservation of angular momentum has one further consequence for spinning. If you swing an arm in one direction, with no external torques, to conserve angular momentum your body will rotate in the opposite direction. The magnitude of this movement is not large enough to impact diving or gymnastics significantly, but falling cats use this fact to manipulate their bodies into comfortable landing positions.

Exercises

In these exercises, (T) refers to thinking problems, conceptual problems requiring no calculations. (C) refers to problems requiring calculus or significant computer calculations. (P) refers to projects; these are ideas for further investigation (hints and resources are at the book's web site).

2.1 An object starts at an angle of θ_0 and completes one lap every T seconds. Find a formula for the angle θ at time t if the object rotates (a) counterclockwise; (b) clockwise.

2.2 An object rotates at n Hz, meaning n laps per second. Compute the object's rotation rate in rpm and its frequency in rad/s.

2.3 Repeat Example 2.2 with a 20% increase in both length and angle.

2.4 (a) A racket is accelerated from rest with angular acceleration 4 rad/s^2. Find the time needed to rotate an angle of π rad, and find the angular velocity at that time. (b) Repeat part (a) with an angular acceleration of 5 rad/s^2. Compare the percentage increases in angular acceleration and velocity.

2.5 Two softball pitchers create an angular acceleration of 15 rad/s^2 on a ball that is initially at rest. If one rotates the ball through an angle of π and the other through an angle of $3\pi/2$, compare the angular velocities of the balls. Speculate on whether a constant angular acceleration throughout the pitching motion is realistic.

2.6 (a) A BMX biker attempting a triple tail whip launches into the air at an angle of 45 deg and speed 30 mph. Ignoring air resistance, how much time in the air does the biker have? After accounting for air drag and launch and landing preparation, suppose there is 1.2 s for the bike to complete its three rotations. What is the average angular velocity during the spins? (b) To do a heelflip, skateboarder Bucky Lasek must rotate his body 180 deg in 0.6 s. Find the average angular velocity during the spin.

2.7 Repeat Example 2.3 with a larger force of 120 lb. Discuss the practical significance of an increased angular velocity.

2.8 A golf club strikes a golf ball 0.5 in below its center of mass with a force of F lb for 0.0005 s. Find the value of F needed to produce a spin rate of 7000 rpm.

2.9 In example 2.3, if the defender hits the runner in a line through the runner's center of mass, then there is no torque applied to the runner. In this case, we compute *linear momentum* to see what happens. Linear momentum is given by $p = mv$. If the runner has velocity 20 ft/s and mass 6 slugs, and the defender has velocity -16 ft/s (the negative indicates a direction opposite that of the runner) and mass 9 slugs, who has the larger momentum? The combined mass of the runner/defender at contact is 15 slugs. Assuming that the total momentum after the contact is the same as before, what is the velocity of the runner/defender combination after contact? The runner has a higher speed, but the defender is larger; in this case, who wins the battle?

2.10 In the situation of exercise 2.9, suppose that a defender has twice the mass of the runner. How fast must the runner be moving to win the momentum contest? Conservation of linear momentum, used in exercise 2.9, assumes that there are no additional forces. Discuss forces that the runner and defender can apply to overcome a deficit of linear momentum.

2.11 Compare the MOI and angular acceleration of a mass of 3 kg rotating 3 m from the center of rotation to a mass of 2 kg rotating 4 m from the center. At what distance would the 2 kg mass need to be to have the same MOI?

2.12 Repeat Example 2.5 with radii of 13 and 14 cm. Explain why the ratio is different here than in Example 2.5.

2.13 A disadvantage of the larger tennis racket could be weight. Assuming a constant mass density of ρ kg/cm^2, and using $m = \rho A$ where A is the area of the racket frame, compare the masses of the rackets in Example 2.5. What would a manufacturer need to do to have the two rackets have the same weight?

2.14 For a given torque, compare the times needed to complete three somersaults in the pike position and two somersaults in the layout position.

2.15 A diver starts spinning in the layout position and then changes to the tuck position. Compare the spin rates.

2.16 To move in a circle of radius r at constant speed v requires an acceleration (to change direction, not speed) of constant magnitude $\frac{v^2}{r}$. The Indy car track in Newton, Iowa, is 0.875 miles long per lap. If the track is circular and the driver maintains a speed of 180 mph, find the magnitude of the acceleration in the form $32c$ ft/s^2 for some constant c. Since 32 is the magnitude of the acceleration due to gravity, 32 is sometimes called 1 g. So c is the number of g's the driver experiences. (The real track is not circular, and is banked; sensors show maximum g-forces of 5.5 at Newton.)

2.17 (T) Two runners have different builds but the same total mass. One has slim lower legs and powerful upper legs, while the other has powerful lower legs and slim upper legs. In terms of rotations about the leg socket, explain why the first runner has the smaller MOI. In fact, long-distance runners tend to have slim and relatively short lower legs.

2.18 (T) Over the years, elite golf and tennis players have become much taller. In terms of club or racket speed, explain this trend.

2.19 (T) Discuss why there might not be constant angular acceleration in a golf downswing. If constant angular acceleration is an invalid assumption, why would we make it?

2.20 (T) Compare the MOIs of a football for a spiral and an end-over-end kick. In 1900, a football was more like a modern rugby ball, much closer to spherical than today's football. Discuss how the modern ball makes passing easier (include MOI as one aspect of your discussion).

2.21 (T) Golf ball manufacturers can alter the MOI of a golf ball. Spin is important in golf to produce extra height and distance, and to have balls land softly, but too much spin could cause a ball to go more up than out, losing distance. Discuss advantages and disadvantages of having a higher MOI golf ball.

2.22 (T) The perimeter weighting system for a tennis racket (see page 31) puts extra weight on the sides (long axis) of the racket. The racket is swung in an upright position for serves, and sideways for ground strokes. For each stroke, discuss how the extra weight improves performance (or not) in cases where the player hits the ball off-center in (a) the left-right horizontal direction; (b) the up-down vertical direction.

2.23 (T) Suppose the defender in Example 2.3 contacts the runner in the chest. In terms of torque, explain why this is less effective than grabbing an ankle.

2.24 (T) To increase MOI, an object's mass can be increased or the distance of its mass from the center of rotation can be increased. Discuss the relative impact of changing mass and changing distance in changing MOI.

2.25 (T) Golfers used to place metal tape around the edge of the backs of their irons. Explain why this could improve performance on off-center hits. (Look up pictures of old "muscleback" and newer "cavity back" irons to see how manufacturers responded.) Discuss where metal tape should be placed to improve the performance of a golf putter.

2.26 (T) One of the classification criteria for snowboard tricks is whether the rider grabs the board or not. In terms of MOI, does grabbing the board make any difference?

2.27 (T) In midair, an athlete swings her right arm across her body to her left shoulder. By conservation of angular momentum, which direction will her body rotate in response?

2.28 (T) Skater Brian Boitano had a signature jump (the "Tano triple") in which he raised his left arm straight into the air while executing a triple rotation. In terms of MOI, does a raised arm make the jump easier or harder?

2.29 (T) In most of the situations we have discussed, larger MOIs are better. Discuss situations in which a smaller MOI is better.

2.30 (C) Assume that a baseball bat has a constant mass density of c slug/in^3. With the bat aligned along the x-axis, it has an outline given by $y(x) = \frac{7}{192}x + \frac{1}{2}$ in for $0 < x < 24$ and $y(x) = \frac{11}{8}$ in for $24 < x < L$ for a bat of length L in. (The barrel of the bat has constant thickness beyond 24 inches.) The cross-sectional area at location x equals $c\pi y(x)^2$ in^2, and the moment of inertia for a swing about the y-axis is $\int_0^L x^2 c\pi y(x)^2 dx$. (a) Show that $I \approx c\pi(0.63L^3 - 2419)$. (b) Show that at a constant angular acceleration α, the bat reaches $\theta = \pi$ with angular velocity $\omega = \sqrt{2\pi\alpha}$ rad/s. (c) From $\tau = I\alpha$, conclude that $\alpha = k/(0.63L^3 - 2419)$ for some constant k. (d) Combine parts (b) and (c) to conclude that $\omega = m/\sqrt{0.63L^3 - 2419}$ rad/s for some constant m. (e) Watts and Bahill measured rotation rates for major league batters and found that $\omega \approx 48 - .34L$ rad/s. Compare the graphs of $f(x) = 48 - .34x$ and $g(x) = 4000/\sqrt{0.63L^3 - 2419}$ for $25 < x < 40$ to see if the two calculations match well. (f) Which assumption(s) in our calculation of ω are invalid?

2.31 (C) Calculate the MOI of a circular tennis racket frame that is 2 cm thick.

2.32 (C) Set up an integral for and then approximate the integral for the MOI of an elliptical racket frame that is 1 cm thick and has (inner) dimensions of 11 cm by 16 cm.

2.33 (P) Find information about the degrees of difficulty for different dives. Do the degrees of difficulty for different positions correspond to the MOIs for those positions?

2.34 (P) A model of a baseball bat is given in exercise 2.30. Modify the model to mimic a hollow aluminum bat. For wooden and aluminum bats of the same dimensions and same weight, how do the MOIs for swinging compare? How do the center of masses compare?

2.35 (P) Explore the usefulness of "corking" a baseball bat. This (illegal) ploy involves hollowing out a cylindrical piece at the end of the bat and filling it with cork (others have used bouncy balls). How much could the MOI change? How much would this increase bat speed? Discuss whether the liveliness of the bat (the coefficient of restitution discussed in Chapter 4) might decrease.

Further Reading

Peter Brancazio's *Sport Science* is an excellent introduction to the physics of sports.

Robert Watts and Terry Bahill's *Keep Your Eye on the Ball* is the best of several physics of baseball books.

A search for "Physics of (favorite sport)" will bring up many results. I can recommend *The Physics of Basketball* (Fontanella), *The Physics of Hockey* (Hache), *Football Physics* (Gay), *Newton's Football* (St. John and Ramirez), *The Science of Soccer* (Wesson), *The Physics and Technology of Tennis* (Brody, Cross, and Lindsey), *Golf Science* (Smith), *The Science of Golf* (Wesson), *Golf By the Numbers* (Minton), and *The Physics of NASCAR* (Leslie-Pelecky).

The Howard Head story is told in Lee Torrey's *Stretching the Limits.*

An impressive figure skating spin (titled "World Record Figure Skating Spin") can be found at https://www.youtube.com/watch?v=AQLtcEAG9v0 accessed July 15, 2015.

Further suggestions can be found in the notes at the *Sports Math* web site.

Chapter 3

Sports Illusions

Introduction

We have all experienced it. Trying to kick or catch or hit a ball, we miss. And then we hear the well-intended but condescending voice of the coach: "Keep your eye on the ball." As we will see, in the most literal of terms this advice cannot be heeded. That is not to say that the advice is bad. Maintaining eye contact for as long as possible is helpful, but it turns out that we humans are not equipped with enough visual acuity to track balls well.

"Keep your eye on the ball" is one of several mythbusting situations that we explore in this chapter. We also try to answer the following questions. Are softball pitchers better than baseball pitchers? Are great athletes born, and not made? Should your football team take a penalty to improve the angle on a field goal? Are referees fully objective and accurate judges? Do high jumpers actually clear the bar? The myths in this chapter are all results of sports illusions, shortcomings in the way our brains process images.

You Can't Keep Your Eye on the Ball

A baseball pitcher throws a fastball across home plate. The batter, following time-honored advice, tries to keep his eye on the ball. (Note that the exact same situation occurs in tennis, softball, and other sports.) The ball, batter's eye, and the middle of home plate form a triangle that changes as the ball moves toward home plate. Assuming that the batter stands 2 feet from the plate, viewed from above the triangle looks like Figure 3.1.

The angle θ can be used to track the batter's gaze as he follows the ball. If θ changes rapidly, then the batter's eye must move rapidly to keep pace. Using basic trigonometry, $\tan\theta = \frac{x}{2}$, where x is the distance between the ball and the plate. Calculus shows that the rate of change of the angle θ (in calculus, we call it the derivative of θ and denote it by θ') is given by

$$\theta' = \frac{2s}{4 + x^2}$$

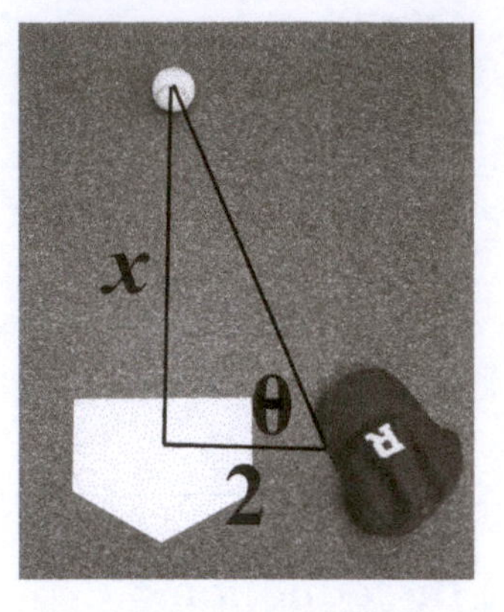

FIGURE 3.1: Eye on Ball

where s is the speed of the ball in ft/s. (Technically, the rate should be negative, since the angle decreases. We are interested in the size of the eye movement, not its direction, so we take the absolute value of rate.) We can now do a simple calculation.

Example 3.1 For a 90 mph fastball, find the maximum value of θ', the rate at which the batter's eyes must move to watch the ball.
Solution. We first convert 90 mph to ft/s, so that the pitch speed is $s =$ (90 mph) (5280 ft/mi)/(3600 s/hr) = 132 ft/s. Our calculus equation then gives us $\theta' = \frac{264}{4 + x^2}$ rad/s when the ball is at position x. What is the largest that this can be? It's clear that the smallest that x^2 can be is 0, occurring at $x = 0$ (when the ball reaches home plate). So the smallest that the denominator can be is 4. The smaller the denominator, the larger the number, so the maximum value of θ' is $\frac{264}{4}$ rad/s = 66 rad/s. This occurs at $x = 0$.

At this point, you may be underwhelmed, but hang on - there's a great punch line coming. Before getting there, let's do a quick reality check on Example 3.1. Does it make sense that the maximum rate of change occurs at $x = 0$? Think about keeping your eye on a race car down the straightaway. You have to move your eyes quickly to follow it as it nears you, but you will have to move your eyes fastest when the car is even with you. The conclusion in Example 3.1 makes sense.

The importance of Example 3.1 depends on the following physiological fact: humans cannot accurately track objects that require angle changes at rates of 3 rad/s or more. The 66 rad/s in Example 3.1 is completely out of the question! In other words, **keeping your eye on this ball is impossible.** (Actually, it might be possible to be looking in the right direction at all times; even so, what you would "see" would be an out-of-focus blur far worse than the ball in the picture at the beginning of the chapter.)

What do major league batters do about this? Watts and Bahill found two strategies, both starting with a smooth tracking of the ball as far as possible.

Some batters then continued moving their eyes at a constant rate, falling behind the ball. Others moved their eyes forward to the place where they predicted that the ball would cross the plate. There is some evidence that excellent hitters are able to track pitches farther than poor hitters. So, "keep your eye on the ball as long as possible" is good advice.

Players now have a variety of video games and training machines designed to improve the players' ability to track the ball and identify clues as to which pitch is coming. The red seams on a baseball have characteristic appearances for pitches, such as a red dot appearing on a slider and apparent train tracks for two-seam fastballs and curveballs. As we will see, this type of visual training can be critical.

You Can't Touch This

The following simple calculation leads to an obvious conjecture. Surprising theories of athletic excellence follow when we find that the conjecture is wrong.

Example 3.2 Ignoring air drag, compare the times for the following two pitches: (a) a major league fastball at 95 mph from a distance of 60 ft; (b) a women's fastpitch softball pitch of 65 mph from 43 ft.
Solution. These are basic "time equals distance divided by rate" calculations, but we do need to convert units. We have 95 mph (5280 ft/mi) $\div$ (3600 s/hr) $\approx$ 139.3 ft/s for (a) and 65 mph (5280 ft/mi) $\div$ (3600 s/hr) $\approx$ 95.3 ft/s for (b). The times are then (a) 60 ft $\div$ (139.3 ft/s) $\approx$ 0.43 s and (b) 43 ft $\div$ (95.3 ft/s) $\approx$ 0.45 s.

Arguments can be made about whether a 95 mph fastball is rarer than a 65 mph softball pitch, and we have seen that air drag changes time of flight significantly. However, the point is that the amounts of time available for batters in major league baseball and women's softball are nearly equal. It would then be an obvious conclusion that major league batters should have no trouble hitting softball pitchers (especially since softballs are larger than baseballs). David Epstein's book *The Sports Gene* describes what happened when this was put to the test. In 2004 and 2005, USA Olympic pitcher Jennie Finch recorded a series of batting practice encounters with baseball's best hitters. Mike Piazza was blown away;

Albert Pujols could not get his bat on a ball. Barry Bonds was almost too befuddled to swing. Alex Rodriguez watched five warm-up pitches and left with the remark, "No one's going to make a fool out of me!" (Which turned out to be not very accurate.)

If the reaction times are the same, then why could the players not hit Finch? An essential part of being a successful batter is to increase the reaction time beyond the 0.4 seconds or so of a pitch. As Finch proved, that is not enough time to see and react. Instead, over time batters develop an unconscious catalog of clues that enable them to predict the speed and location of pitches before they leave the pitcher's hand. Given no catalog of clues to Finch's delivery, even the best and quickest hitters were hopeless, to the point of almost never even getting a bat on the ball.

Some clever experiments have shown that these clues exist. If you are shown a video of a pitch that stops right before the ball is released, could you predict the speed and location of the pitch? The better the batters, the more accurate their predictions are. Can we determine which clues are being used? Various *occlusion* experiments doctor the video to black out, for instance, the pitcher's shoulders. If the accuracy of your predictions decreases, then there is evidence that you are using something about the motion of the shoulders.

You Can't Teach Size

The preceding discussion indicates that the experience and practice of building the catalog of clues is more important than the raw speed or visual acuity of the athlete. Questions of how much of player success is genetic and how much is due to environment (including hard work) are intriguing but hard to answer.

An important addition to the Jennie Finch study, also reported in *The Sports Gene,* is that in fact the top baseball players and softball players have outstanding eyesight. Actually, "outstanding" is an understatement, as major leaguers and 2008 U.S. Olympians tested at an average visual acuity of 20/11, with superior depth perception and contrast sensitivity to boot. Of the Los Angeles Dodgers tested, about 2% scored a 20/9 acuity rating, considered to be near the limit of human capacity. The top hitters are equipped with outstanding genetics as well as well-developed brains.

There are no simple answers. Almost nobody succeeds in sports without long hours of dedicated practice (or, if you're Allen Iverson, game action). You can't hit Jennie Finch without many reps against her or similar softball pitchers. An oversimplification that appears in many sound bites is the 10,000 Hour Rule. This says that top performers in almost any field need to have logged at least 10,000 hours of deliberate practice before reaching elite status. The term "deliberate" encompasses much. Your practice can't be dull and plodding, just putting in the hours; it must be focused practice that actively addresses any weaknesses that you may have. The easiest way to characterize deliberate practice is to say that this is what someone completely dedicated to becoming the best of all time would do. Unfortunately, this makes our definition circular.

Contrary to a common misinterpretation of the 10,000 Hour Rule, putting in the hours does not guarantee success. When I was a college student, putting in 10,000 hours at a basketball camp teaching low-post skills would not have made me an NBA-caliber player. My (mostly) genetic endowment of 5'11" height and no jumping ability precluded that. However, all of that practice would have helped me abuse the other students in the pick-up games in the gym.

An interesting question in the sports version of "nature versus nurture" is whether hard work, persistence, or love of the game have a genetic component. Sons of football players often love football, but that could be easily explained by an environment of positive experiences with football. A related issue involves the effect of training. Some people respond dramatically to training, increasing strength or speed in leaps and bounds. For others, the same training produces no improvement. Is "trainability" genetic? The rapid conversion of Usain Bolt from a high school cricket and football player to the world's fastest sprinter ever seems to defy the 10,000 Hour Rule. Bolt responded to standard training techniques with spectacular improvements in time. Is his greatest gift a speed gene or a trainability gene? Either way, his accomplishments are remarkable.

You Can't Afford the Yardage

A football player is driven out of bounds just two yards from the goal line, bringing up fourth down. The coach calls for a field goal, after taking a 5-yard penalty to "improve the angle." What is wrong with this picture? You might first object to the conservative choice of trying a field goal; we explore the many pros and few cons of going for it in Chapter 9. From the standard television camera behind the goalposts, the angle for a short field goal from a hash mark does look severe. But, does it really increase the angle to move back 5 yards?

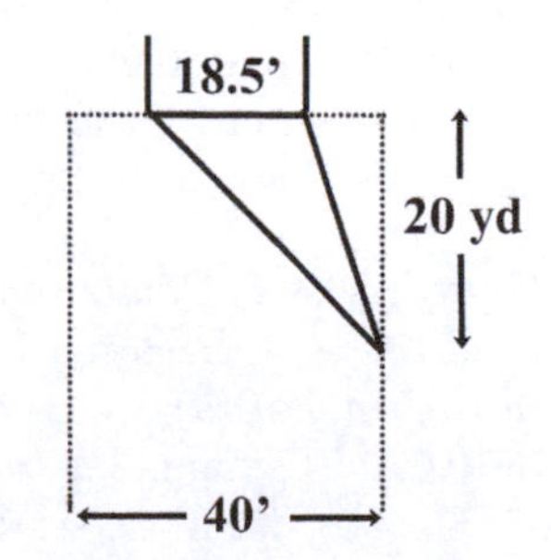

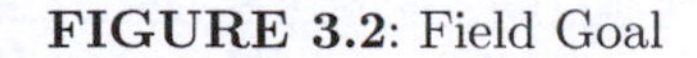
FIGURE 3.2: Field Goal

Example 3.3 The goal posts in college football are 18 ft 6 in apart and the hash marks are 40 ft apart. Compare the angle for a field goal from the right hash mark from distances of (a) 20 yards and (b) 25 yards.
Solution. With the ball on the right hash mark, draw a triangle with vertices at the ball, the place where the hash mark would hit the back of the end zone, and the far goal post. (See Figure 3.2.) This is a right triangle with sides of 20 yards (in part (a)), 29 ft 3 in (20 ft from the hash mark to the center of the goal post plus 9 ft 3 in more), and the length of the hypotenuse (which we will not use). In a common unit of feet, the sides are 60 and 29.25. The angle θ at the ball satisfies $\tan\theta = \frac{29.25}{60}$ and so $\theta = \tan^{-1}\left(\frac{29.25}{60}\right) \approx 0.4536$ rad $=$ 0.4536 rad $\times$ $180/\pi$ deg/rad ≈ 25.99 deg. To compute the angle to the near goal post, we only change the distance from the hash mark to the near post to $20 - 9.25 = 10.75$ ft. The angle changes to $\theta = \tan^{-1}\left(\frac{10.75}{60}\right) \approx 0.1773$ rad $=$ 0.1773 rad $\times$ $180/\pi$ deg/rad ≈ 10.16 deg. The margin of error for the kick, the angle between the two posts, is approximately $25.99 - 10.16$ deg $= 15.83$ deg. For part (b), we change the two calculations in part (a) by replacing 60 with 75. The angle between the two posts is now $\tan^{-1}\left(\frac{29.25}{75}\right) - \tan^{-1}\left(\frac{10.75}{75}\right) \approx$ 0.2295 rad ≈ 13.15 deg, 2.7 degrees less than the shorter kick.

Moving 5 yards from a 20-yard field goal to a 25-yard field goal reduces the angle for a successful kick by nearly 17%. Why have teams thought for years that this was good strategy? An argument can be made that the angle is not the only variable to take into account. From the middle of the field, the kick is perpendicular to the line of scrimmage and directly into the crowd behind the goal post; numerous visual cues help you line up the kick. From the hash mark, these cues are largely gone, and lining up may be more difficult. If there is a decrease in accuracy due to the odd look of the short kick, perhaps that makes up for the loss in accuracy due to a decreased angle of success.

A different situation sometimes occurs at the end of football games. A team running the clock down before trying a game-winning field goal may run one last play where the quarterback takes the snap, slides over to the center of the field, and kneels. The play may lose a yard or two, but getting the ball into the center of the field is considered worth it. From the perspective of angles, is this true?

Example 3.4 Compare the angles for a college football (a) 35-yard field goal from the right hash mark and (b) 37-yard field goal from the center.
Solution. For (a), we repeat the analysis from Example 3.3 with 60 ft replaced by 105 ft. The angle is $\tan^{-1}\left(\frac{29.25}{105}\right) - \tan^{-1}\left(\frac{10.75}{105}\right) \approx 9.72$ deg.

For (b), we must start over. Consider the two triangles to be drawn, one to the goal post on the left and the other to the goal post on the right. In each case, the leg of the triangle along the goal line has length 9.25 ft. Thus, the angles to the two posts are the same; the angle of success is twice this angle, or $2\tan^{-1}\left(\frac{9.25}{111}\right) \approx 9.53$ deg. In this case, dropping back two yards decreases the angle more than centering the ball increases the angle.

You will discover in exercise 3.8 how much centering the ball improves the angle, and how far back the quarterback can kneel before the overall angle is decreased.

It should be noted, and will be explored in the exercises, that college goalposts and hashmarks are wider than those in the NFL, and narrower than those in high school. The numerical conclusions drawn will be different with different field specifications, and the interpretations vary with the differences in abilities of the kickers.

You Can't Bend That Way

In Chapter 2, we talked about the advantages of a long swing in golf or tennis. I once played against a golfer who essentially had no backswing; he raised his club into a good position and then started his downswing. Why do golfers take vigorous backswings, and pitchers use complicated windups? The picture to the right gives a clue.

FIGURE 3.3: Club Bend

In Figure 3.3, the clubhead has stopped going back but has not yet started forward. The clubhead's velocity, then, is zero. So why not use my opponent's strategy and simply place the club in that position? The answer is that you *could* place the clubhead in that location, but you would not get the bending of the club. The bending/unbending of the club works enough like a spring that it is worth a quick detour to discuss energy in a spring.

Example 3.5 Suppose a spring is compressed by 2 m and then released. Track its energy as (a) it moves through its resting position and (b) reaches its maximum stretching position.

Solution. At the beginning of this process, the spring has zero kinetic energy ($K = \frac{1}{2}mv^2$). However, after its release it picks up speed and therefore kinetic energy. In its initial position, we say that the spring has potential energy U to indicate that it has the potential for an increase in kinetic energy. (a) A spring at its resting position has zero potential energy; if it starts with no kinetic energy, it will remain stationary. However, in our situation it has accumulated kinetic energy. So, $U = 0$ and $K > 0$. At position (b), the spring has stopped stretching and not yet started to compress; therefore, it has no kinetic energy. However, it now has potential energy (the spring will compress and regain kinetic energy). In the absence of friction, conservation of energy tells us that the sum $K + U$ will remain constant, so that the potential energies at the beginning and at position (b) will be the same, with both equal to the kinetic energy at position (a).

In Figure 3.3, the golf club has little kinetic energy, but the bending of the club gives it potential energy; like a compressed spring, it will snap back to its resting position, creating kinetic energy. This potential energy translates into extra clubhead speed, and is created by the transition from backswing to downswing. This is an important advantage of an energetic backswing.

Similarly, a pitcher's elbow undergoes unnatural stresses as the pitcher transitions from arm cock to follow through. The following example gives an idea of how much force is applied.

Example 3.6 Research indicates that up to 120 N-m of valgus torque (considered the prime cause of elbow injuries in pitchers) is exerted on a major league pitcher. At a distance of $\frac{1}{3}$ m (about the distance from hand to elbow), how much force is needed to create this torque?
Solution. Recalling that $\tau = FL$, we want 120 N-m $= (F)(\frac{1}{3}$ m) so that $F = 120 * 3 = 360$ N. For comparison purposes, 360 N is about 81 pounds, enough force to create significant potential energy in the arm to increase throwing speed.

You Can't Make That Call!

Sports referees have difficult and typically thankless jobs. One of the hardest referee decisions in sports is the offside/onside call in soccer/football. The referee must keep multiple players and positions in mind to make the correct call. Research showed that in the 2002 World Cup mistakes were made on offside calls an astonishing 26% of the time. Improved research and training reduced that percentage to under 10% in the 2006 World Cup.

In the most common scenario, a player passes the ball ahead to a teammate (an attacker). At the instant the ball is passed, imagine stopping the action and drawing imaginary lines through each player and parallel to the goal line. Typically, the goalkeeper's line will be closest to the goal line. The next closest line through a defender (called the last defender) is the offside line. If the attacker's line is closer to the goal line than the offside line, then the attacker is offside, the play is stopped, and the

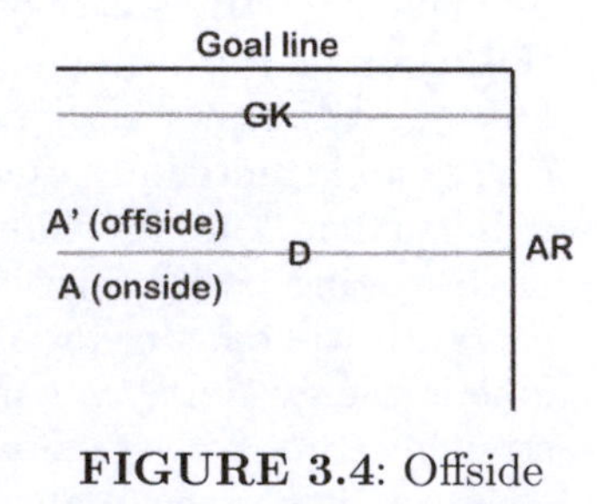

FIGURE 3.4: Offside

defending team is awarded a free kick. In Figure 3.4, an attacker at position A is onside, while an attacker at position A' is offside. The AR (assistant referee, or sideline referee) is tasked with monitoring which defender is the last defender and whether attackers are onside or offside.

There are variations on the rule, but we will work with this basic situation and examine three possible ways for the AR to unknowingly make an incorrect call.

The first possibility to consider is the shift-of-gaze error. If the AR is watching the ball until it is kicked, and then locates the attacker and last defender, during the time required for the AR to re-focus the attacker could move from position A (onside) to position A' (offside). Attackers try to time their run so that they pass the last defender shortly after the ball is kicked, giving them a head start on running to the pass. As seen in Example 3.7, the referee could see the attacker as being offside when, at the exact instant of the pass, the attacker was onside. However, research shows that elite referees always watch the attackers and defenders, tracking the ball only through peripheral vision. At the highest level, shift-of-gaze errors rarely occur.

A second possibility is optical error. Research has shown that ARs are only rarely positioned on the offside line. They are typically a meter or two closer to the goal than the offside line. Because the AR's line of sight is not parallel to the offside line, mistakes such as the one in Example 3.7 can be made. In studies, slightly less than half of the offside mistakes are consistent with optical error. A higher percentage of errors can be explained by the flash-lag effect, to be discussed after a quick example.

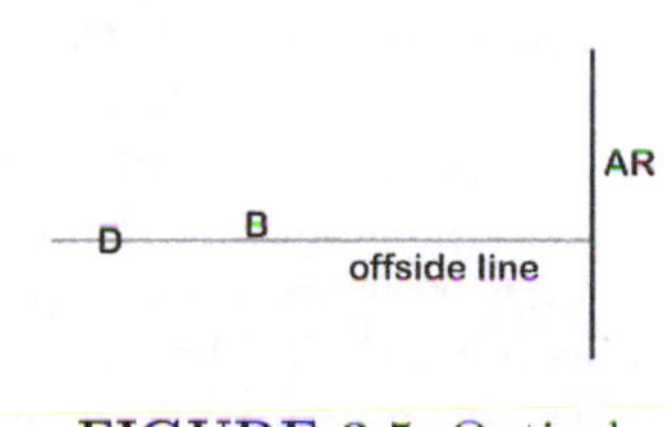

FIGURE 3.5: Optical

Example 3.7 Show that incorrect calls are made in each of the following situations. (a) (Shift of Gaze) The attacker in Figure 3.4 is onside at position A, one-half meter behind the offside line and running at a pace of 10 m/s; the defenders are not moving. If it takes the AR 0.1 seconds to shift gaze from the ball being passed to the attacker, will the AR make the correct call? (b) (Optical) The attacker in Figure 3.5 is offside at position B, 0.3 m in front of the offside line and 30 m from the touchline (sideline). The last defender is 40 m from the touchline, and the AR is standing 2 m in front of the offside line. Will the AR make the correct call?

Solution. (a) At a speed of 10 m/s, the attacker will advance 10(0.1) m = 1 m in 0.1 seconds. The attacker has moved from 0.5 m behind the offside line to 0.5 m in front of the offside line, and will be called offside.

(b) Think of a coordinate system where the defender is the origin and the offside line is the x-axis. The line from the defender to the AR has slope $\frac{2}{40} = 0.05$ and has equation $y = 0.05x$. This is the AR's effective offside line; everybody below this line will be judged onside. The attacker is 10 m closer to the touchline, and thus is at $x = 10$. At $x = 10$, the offside line passes through $y = 0.05(10) = 0.5$. The attacker at 0.3 m is below this line, and thus appears onside to the AR.

We have seen two ways for referees to miss this difficult call. A physiological phenomenon called the *flash-lag effect* presents another challenge. In the television series *Brain Games*, this effect is presented as a game in which a football (American style) moves at constant speed across the screen. At some time, a red dot flashes in the upper-right corner; we viewers are supposed to determine the position of the football when the flash occurs. The interesting result is that 95% of test subjects misplace the football, believing that it has gone farther across the screen than it has. The explanation is that it takes about one-tenth of a second for the brain to process the surprise flash, and the brain marks the location of the football after that processing occurs. As in Example 3.7, during this time lag the football moves to a new location, which we incorrectly "see" as the location at the time of the flash.

Substitute moving football players for a moving football and a kick for a red dot and you have the main components of the offside call. Interestingly, research has shown that good referees make mistakes that are consistent with flash-lag effect, but elite referees often do not. They have learned some way to correctly compensate for the flaw in their human vision processing systems.

An interesting variation of this result occurs in baseball. The play in question is at first base, when a throw from an infielder pops into the first baseman's mitt at the same time as the runner's foot hits the bag. Is the runner safe or out? I've always been amused at the coolness of the umpires, who will look at the play and eventually raise a fist casually to indicate that the runner is out (in fact, *clearly* out, don't even think about arguing).

FIGURE 3.6: Out?

Research shows that umpires raise their fists too often, calling runners out who are actually safe at first. The reason for this is rooted in how the human brain processes sight and sound. Major league umpire Mike Winters describes an umpire teaching school technique of blindfolding umpires. This is not to validate the loud fan in the thirtieth row who thinks a blind person could make more accurate calls. This is to teach umpires to listen for the pop of the ball in the mitt. Winters says that umpires who listen for the ball pop and the thud of the runner's foot on the bag never miss the call, but those who try to visually determine whether the ball reaches the mitt before the foot touches the bag often make mistakes.

Umpires are taught to keep their eyes on the base while listening for the ball to hit the mitt. This way they avoid the shift-of-gaze and flash-lag errors discussed above. However, this runs the umpire straight into a different cognitive illusion. Over 100 years ago, German experimenter Wilhelm Wundt discovered that when an audio event is paired with a visual event, the brain system that syncs the two tends to pre-date the visual event. Thus, if the ball pops into the glove at the same time as the foot hits the bag, the brain backs the runner up a small amount and syncs that previous position with the sound. The umpire perceives the sound as occurring before the runner hits the bag, and makes the out call.

There is a clear advantage for the brain to work this way. Think about a door being slammed from a long distance away. The sight of the door closing reaches your eyes before the sound of the door shutting reaches your ears (light travels faster than sound). An intelligent brain backtracks the visual image to correctly sync sight and sound. Unfortunately, this brain feature becomes a bug when trying to make the correct call at first base.

On June 2, 2010, Detroit pitcher Armando Galarraga found out that the Wundt effect is not always active. Jason Donald hit a ground ball to Detroit first baseman Miguel Cabrera, who flipped the ball to Galarraga, beating Donald to the base for the out ... except that umpire Jim Joyce called the runner safe. The tragedy is that an out would have been the 27th and final out of a perfect game.

Think about Joyce's situation: with the Detroit crowd screaming for a perfect game, he had to *hear* a soft flip from Cabrera hitting Galarraga's glove. In fact, Galarraga snagged the ball in the webbing of his glove, so there was likely no noise at all. If Joyce, known as an excellent umpire, had to rely on eyesight only, the call was made difficult by all of the factors that afflict assistant referees in soccer. The bottom line was a very bad bit of luck for Galarraga, although the extra publicity for the bad luck and widespread praise of the great sportsmanship shown by both Galarraga and Joyce may have made up for the loss of a perfect game.

You Can't Clear That Bar

The final example in this Chapter (more to follow in the exercises, though!) involves high jumping. If you are old enough, you may remember the straddle technique that was routinely used until the 1970s. If not, search for an old video; the technique looks very odd to modern eyes. High jumpers now use a technique introduced to the world by Dick Fosbury in the 1968 Olympics (and therefore known as the Fosbury Flop) where they jump backward, arch their backs and then pull their legs over the bar. A quick look at the winning jumps in various Olympics (Figure 3.7) suggests that success with the straddle technique had plateaued by 1976, and the flop technique took it to a new level, at which we are now stuck. (To be fair, a plot of world record jumps tells a different story.)

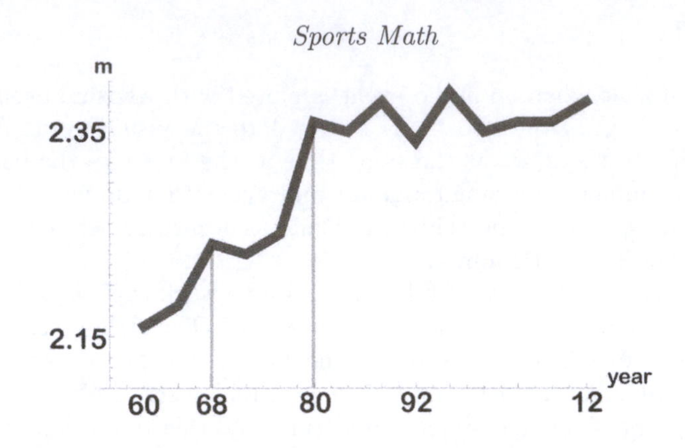

FIGURE 3.7: Heights of Winning Men's High Jumps, Olympics, 1960-2012

FIGURE 3.8: High Jump

Here's the amazing part: in terms of the Newtonian mechanics outlined in Chapter 1, the best high jumpers do not clear the bar! To be precise, the center of mass of the high jumpers can pass underneath the bar on a successful jump (recall that our equations in Chapter 1 only track the center of mass). How can this be? Take a careful look at the jumper in Figure 3.8. Her head has cleared the bar and is now underneath the height of the bar, as are her shoulders, lower legs and feet. More of her mass is below the bar than above, and her center of mass (where the sums of mass times distance for her body parts balance) is below the bar. She has performed a magic trick, clearing a bar that is higher than she can jump.

Exercises

In these exercises, Ⓣ refers to thinking problems, conceptual problems requiring no calculations. Ⓒ refers to problems requiring calculus or significant computer calculations. Ⓟ refers to projects; these are ideas for further investigation (hints and resources are at the book's web site).

3.1 Rework Example 3.1 with a 75 mph curve ball.

3.2 In Example 3.1, (a) find the maximum ball speed at which a human could track the pitch all the way to home plate; (b) determine how far the batter can track the pitch (i.e., the smallest x such that the rotation rate is 3 rad/s or less).

3.3 A tennis player tries to track a serve that is on a line 2 feet to the side at the speed 105 mph. (a) What is the rotation rate needed to track this serve all the way to the racket? (b) Given a maximum tracking rate of 3 rad/s, how far can the player track the serve? (c) Can a linesperson sitting 30 feet to the side track the ball?

3.4 A spectator sits 300 feet away from a racetrack and tries to follow a race car moving at 180 mph. (a) What is the rotation rate needed to track the car all the way? (b) Given a maximum tracking rate of 3 rad/s, what is the closest the fan can sit to the track and completely track the car?

3.5 Ignoring air drag, find the reaction times for the following: (a) a tennis serve at 130 mph from 78 feet away; (b) a lacrosse shot at 100 mph from 30 feet away; (c) a penalty kick at 80 mph from 12 yards away; (d) a hockey shot at 90 mph from 30 feet away; (e) a line drive at 150 mph at a third baseman 85 feet away.

3.6 Data from MLSsoccer.com indicates the following breakdown of birth months for United States U-17 and U-20 national team players: 14.1% (Jan), 10.8% (Feb), 12.0% (Mar), 9.9%, 8.5%, 7.0%, 6.0%, 7.7%, 6.0%, 7.1%, 4.5% (Nov), 6.5% (Dec). If all birth months were equally likely, what percentage would each month have? What is the average absolute difference between expected and actual percentages? "U17" means players must be under the age of 17 as of January 1. Compare the ages of the oldest U17 player and the youngest (of those who would not qualify for U16). Does birthdate matter?

3.7 Repeat Examples 3.3 and 3.4 for (a) high school dimensions of goal posts 23 ft, 4 in apart and hash marks 53 ft, 4 in apart; (b) NFL dimensions of goal posts 18 ft, 6 in apart and hash marks 18 ft, 6 in apart.

3.8 In Example 3.4, (a) how many yards lost could a team absorb and still have an equal angle for kicking? (b) Repeat for high school kickers.

3.9 At the top of a pole vault, a vaulter of mass m kg at height h m has gravitational potential energy of mgh J. The kinetic energy of the vaulter before planting the pole is $\frac{1}{2}mv^2$ where the vaulter's running speed is v m/s. Assuming that these are equal, and that the top running speed of a vaulter is 12 m/s, find a quick estimate of the maximum height of a pole vault. The world record in 2014 is 6.16 m; how does your estimate compare?

3.10 In Example 3.7 (a), if the attacker is onside by 1.5 m and the defender is also running at 10 m/s but in the opposite direction, show that an AR with a 0.1 s gaze shift will get the call wrong.

3.11 In Example 3.7 (b), determine the set of positions (distance on- or offside and distance from the touchline) in which (a) the AR incorrectly sees an offside attacker as being onside; (b) the AR incorrectly sees an onside attacker as being offside.

3.12 Repeat exercise 3.11 if the AR is only 1 m ahead of the offside line.

3.13 The definition of offside involves the location of the attacker at the time of contact with the ball. If this contact lasts 0.05 s, determine circumstances in which the attacker could be legally both onside and offside.

3.14 Plot a graph similar to Figure 3.7 but using world record high jumps instead of Olympic winning heights. In your new graph, does 1968 appear to be an important year? Which graph, yours or Figure 3.7, do you think more accurately reflects the evolution of the high jump? Give reasons.

3.15 (T) Baseball batters talk about "rising fastballs" and curve balls "dropping off the edge of a table." For a batter who redirects his eyes to where he predicts the pitch will end up, explain how this illusion could occur.

3.16 (T) Ted Williams, considered by many to be the greatest hitter ever, claimed that he could see the bat hit the ball. His eyesight, measured in standard terms to be 20/9, was extraordinary but do you think he could track the ball all the way to home plate? If not, would it be possible for him to clearly see the contact of bat and ball?

3.17 (T) Submarine-style pitcher Chad Bradford threw underhanded, sometimes scraping his fingers on the ground. Even though Bradford did not throw hard (85 mph or so), explain why batters might have trouble hitting his pitches.

3.18 (T) Michael Jordan famously took time off in the middle of his NBA career to play professional baseball. Even though Jordan was a fabulous athlete, explain why it would be reasonable to expect him to struggle as a hitter.

3.19 (T) Hold a ruler or meter-stick vertically, have a friend place two fingers on either side of the stick, drop the stick, and have your friend grab it as quickly as possible. Explain how this can be used to determine your friend's reaction time.

3.20 (T) The television show *Sport Science* tested drag racer Hillary Will's reaction time (critical to racers) using the starting sequence of descending lights used in races. She was clocked with a reaction time of 0.001 s. Human reaction times for seeing a green light and pushing a button are more typically on the order of 0.2 s. Explain this apparently superhuman performance.

3.21 (T) You are familiar with the "on your mark, get set" BANG that starts running races. To try to guarantee that sprinters are reacting to the gun and not anticipating the start, there are harsh penalties for false starts and a "reaction time" of less than 0.1 s is considered a false start. Discuss how reaction time might be measured and why 0.1 is the threshold.

3.22 (T) Mo Farah of Great Britain won the 10,000 m race at the 2012 Olympics. The previous four Olympics had been won by Ethiopians. For an event dominated by a country or region, give reasons why the domination might be mostly nature (genetics) and reasons why the domination might be mostly nurture (training and motivation).

3.23 (T) The armspan-to-height ratio for most humans is about 1 (measure yourself!). The average ratio in the NBA is 1.063; while this may not seem much different than 1, a ratio higher than 1.05 triggers tests for Marfan's disease. Discuss the advantages for an NBA player to have a high ratio.

3.24 (T) The human body shows remarkable abilities to adapt to training and improve at specific tasks. We sometimes underestimate the specificity: for example, improvements in lifting free weights do not always transfer to other measures of strength. Discuss the benefits of swinging a heavy bat in warmup or wearing a high-drag suit for swimming training. Explain why these do not necessarily help batting and swimming performance.

3.25 (T) Discuss ways in which a tennis serve motion could bend the server's elbow unnaturally, and why this could increase serving speed.

3.26 (T) If an attacker is running toward the goal, under the flash-lag hypothesis discuss in which situation the AR is more likely to miss an offside call: (a) the defender is stationary; (b) the defender is running away from the goal in an offside trap.

3.27 Ⓣ Test yourself on the *Brain Games* football challenge: season 2, episode 11, *Illusion Confusion*.

3.28 Ⓣ Explain why the tendency of baseball umpires to call runners out at first base when they are actually safe is contrary to the flash-lag theory.

3.29 Ⓣ In Figure 3.6, in 1980 there is a significant increase in winning height. Give a reason that the introduction of the Fosbury Flop in 1968 might explain this increase.

3.30 Ⓒ Given that $\tan\theta(t) = \frac{1}{2}x(t)$ and $x'(t) = s$, derive the rate of change equation $\theta'(t) = \dfrac{2s}{4+x^2}$.

3.31 Ⓒ Generalize exercise 3.30 to the case where $\tan\theta(t) = \frac{1}{L}x(t)$ for some length L.

3.32 Ⓒ Adapt the baseball equations of motion ($x' = v_x$, $y' = v_y$, $z' = v_z$, $v_x' = -f_d v_x\sqrt{v_x^2+v_y^2+v_z^2} + f_m(w_y v_z - w_z v_y)$, $v_y' = -f_d v_y\sqrt{v_x^2+v_y^2+v_z^2} + f_m(w_z v_x - w_x v_z)$, $v_z' = -32 - f_d v_z\sqrt{v_x^2+v_y^2+v_z^2} + f_m(w_x v_y - w_y v_x)$, $f_d = 0.002203$, $f_m = 0.000632$) to model a pitch that starts at the point $(55, 5)$ and heads toward home plate at $(0, 0)$. Assuming pure backspin and an initial velocity that is horizontal at 95 mph, find the spin rate in rpm that would actually produce a rising fastball. (Hint: you need the initial v_z' to be positive.)

3.33 Ⓒ Referring to exercise 3.32, simulate the paths of fastballs that are thrown horizontally (a) at 95 with 1500 rpm backspin; (b) at 100 mph with 1600 rpm backspin. (c) Find the height of each pitch when it is 20 feet from home plate. Would the batter be able to tell the difference? (d) Find the height of each pitch when it reaches home plate. (e) If the batter uses the height 20 feet away to predict the height at home plate, why might he think that the 100 mph/1600 rpm fastball hopped at the end?

3.34 Ⓒ For the kicker in Example 3.3, find the distance x at which the angle is maximized. Why is this not a valid distance in football?

3.35 Ⓒ For kickers in high school (goal posts 23'4" wide and hash marks 53'4" wide), find the distance x at which the angle is maximized. Is this a valid distance in football?

3.36 Ⓒ A hockey player races down the ice in a straight line parallel to the sideboards and 3 feet wide of the net. The net is 6 feet wide. At what point is his left-right margin of error at its maximum?

3.37 Ⓒ A soccer player shoots from 20 yards out. Where should the keeper be positioned to minimize the angle for scoring a goal? The goal is 24 ft wide; assume that the keeper can cover 12 ft. Minimize the error if the shooter is (a) 4 ft wide of the left post; (b) 4 ft inside the post; (c) generalize your answers.

3.38 Ⓒ Generalize exercise 36 to a shooter who is a ft wide of the net, with a net that is b ft wide. Show that the maximum margin of error is at $\sqrt{(a+b)b}$ ft.

3.39 Ⓟ Explore the advantages or disadvantages of being lefthanded in sports. Possible research avenues are studies showing that (1) lefthanded athletes die younger, or (2) lefthanded batters have higher batting averages, or (3) there are a disproportionate number of lefthanders in elite tennis and baseball.

3.40 (P) Investigate one of the following claims: (1) football teams from warm-weather cities do not perform well in the cold; (2) championships are won by teams with strong defenses (not strong offenses); (3) at most universities, college football brings in enough money to pay for non-revenue sports.

Further Reading

Robert Watts and Terry Bahill's *Keep Your Eye on the Ball* is the source for the idea that batters can't keep their eyes on the ball, and for the research of what batters actually do.

David Epstein's *The Sports Gene* is an excellent read on many aspects of the nature versus nurture debate.

Malcolm Gladwell's *Outliers* gives an enjoyable overview of research into elite performances in a variety of fields. K. Anders Ericsson is a proponent of the 10,000 hour rule in sports; see "The Role of Deliberate Practice in the Acquisition of Expert Performance" in *Psychological Review* (1993).

The Sloan Sports Analytics Conference generously posts videos of many of its sessions. A discussion between Epstein and Gladwell on nature versus nurture occurred in 2014, and is titled "10,000 Hours vs. The Sports Gene."

Harold Klawans's *Why Michael Couldn't Hit* gives a wealth of information about the neurological basis of sports performances.

More about umpires making the call at first base can be found at http://www.nytimes.com/2011/06/11/sports/baseball/first-base-umpires-call-them-as-they-hear-them.html accessed 7/23/2015.

Gordon Russell's *Sport Science Secrets* discusses a variety of sports facts and myths, including the close calls at first base and the home field advantage.

Sources for the offside call include "Offside decision making in the 2002 and 2006 FIFA World Cups" by Catteeuw, et al; "The Effects of Additional Lines on a Football Field on Assistant Referees' Positioning and Offside Judgments" by Barte and Oudejans; "Visual Scan Patterns and Decision-Making Skills of Expert Assistant Referees in Offside Situations" by Catteeuw, et al.

"Seeing the Benefit: MLB teams focus on enhancing players' visual training" by Stephanie Apstein discusses players' visual acuity and new training techniques. *Sports Illustrated* April 13, 2015 issue.

Sport Science "Myths" episode with Jason Zuback, https://www.youtube.com/watch?v=Ito3BSO-St8 ; "Long Driver" with Jamie Sadlowski, $https://www.youtube.com/watch?v = oUZeBzkcLU0$ accessed 7-24-2015.

Swimming training articles include $http://www.swimmingscience.net/2013/08/drag - suits - part - ii.html$, accessed 7-24-2015.

The *Sport Science* episode "Lose the Weight" explores the use of a weighted bat, $https://www.youtube.com/watch?v = 0_vR8U_KrhY$ accessed 7-24-2015.

Chapter 4

Collisions

Introduction

Football is a sport of collisions. Fans and players celebrate big hits as enthusiastically as touchdowns. The same is true of hockey, where a strong check always earns a roar of approval from the fans.

The same is true of tennis, ... Did I get your attention? Even a McEnroe-esque screaming tantrum on the tennis court is unlikely to have physical contact, so why would I claim that tennis is a sport of collisions? Here's why: every point starts with a collision between racket and ball, followed by a collision between ball and ground, and so on. The motion of the ball, and therefore the outcome of each point, is determined by these collisions.

In this chapter, we look at collisions of many types. Some are the person-to-person collisions that energize sports fans, but most are collisions between the tools of the sports, the commonplace occurrences that give the sports their distinctive characters. We will see that hockey, in addition to bone-crushing checks, has the most intricate of collisions in the slap shot's dance of the stick and puck. We will find answers to the following questions. How much force does a football player absorb? Why does a kangaroo jump better than a human? Can a baseball player hit a ball farther by gripping the bat more tightly? Where is the sweet spot on a tennis racket? How do sports regulate the equipment to keep the games competitive? Why did Rick Barry shoot free throws underhanded, and is it a coincidence that he is one of the best of all time?

Linear Momentum

In a 2014 college football game, Clemson defender Jayron Kearse tackled Louisville running back Dominique Brown head-on (let's say "shoulder-on" to keep it legal), stopping him in his tracks. The play was unremarkable, except for the fact that Brown weighed in at 230 pounds compared to Kearse's 200 pounds. How is the smaller man able to win the collision? You probably know that the answer has to do with speed, that the impact of a hit depends on both the mass and the speed of the hitter. The follow-up question is to ask for the exact combination of mass and speed (add them? multiply them? square them?) to determine the outcome of a collision.

The answer, as usual, comes from Newton's Second Law. Rather than the form $F = ma$, replace acceleration with change in velocity (Δv) divided by change in time (Δt). Multiplying across by Δt, we have that

$$F\Delta t = m\Delta v = \Delta(mv)$$

where we assume that the mass m is constant. We give names to both sides of this revised Second Law. The combination $F\Delta t$ is called **impulse** (J) and the combination mv is called **linear momentum** (p). Our equation is then the **impulse-momentum equation**

$$J = \Delta p$$

or, in words, impulse equals the change in linear momentum.

When there is no force, the impulse is zero, making the change in linear momentum zero. This is **conservation of linear momentum**:

In the absence of external forces, linear momentum is conserved.

This enables us to solve our football problem. Assume that just before the collision Kearse is moving in one direction, which we will call the positive direction, with speed v_k. His linear momentum is $+m_k v_k$. Brown moves in a negative direction with linear momentum $-m_b v_b$. The total linear momentum of the two players before the collision is $+m_k v_k - m_b v_b$. Assume that there are no external forces (no other players joining the pile, no pushing on the ground by either player, and so on). Then the total linear momentum is conserved, so that the total linear momentum before the collision equals the total linear momentum after the collision. The situation after the collision is a single Kearse/Brown tangle of mass $m_k + m_b$ moving with a combined velocity of v_c. Then

$$m_k v_k - m_b v_b = (m_k + m_b)v_c$$

where Kearse wins (the players move in the Clemson direction) if $v_c > 0$ and Brown wins if $v_c < 0$.

We are almost there. Kearse wins if the total linear momentum is positive. Using the total linear momentum before the collision, this means that $m_k v_k - m_b v_b > 0$, which happens if $m_k v_k > m_b v_b$. So, we have a simple conclusion: **whichever player has the larger linear momentum wins**!

Example 4.1 A 230-pound running back moving at 20 ft/s collides with a 200-pound defensive back. Assuming no external forces, determine how fast the defensive back needs to be moving to stop the running back.
Solution. Since an object with mass m has weight mg for the gravitational constant g, the masses of the two players are $230/g$ slugs and $200/g$ slugs, respectively. Assume that the defensive back runs with speed v ft/s. By the discussion above, the assumption of conservation of linear momentum (no external forces) implies that the player with the larger linear momentum wins. The defensive back stops the runner if $v * 200/g \geq 20 * 230/g$. Multiplying by g removes that constant, and we conclude that $v \geq 20 * 230/200 = 23$ ft/s. The running back weighs 15% more, so the defender needs to be moving 15% faster to stop him.

Impulse and Force

We know how fast Kearse needs to move to stop Brown, but how much force does he deliver? The impulse-momentum equation, which states that impulse equals change in linear momentum, is a place to start. This time, we focus on just the running back Brown, for whom conservation of linear momentum does not apply (since the defender is intent on delivering an external force to him). He enters the collision with linear momentum mv and exits the collision with linear momentum 0 (he has 0 velocity). Therefore, his change in linear momentum equals $mv - 0$. Using the values from the example above, $m = 230/g$ slugs and $v = 20$ ft/s, we have $\Delta(mv) = 4600/g$ slug-ft/s. The awkward unit of slug-ft/s is equivalent to lb-s, which is not coincidentally a unit for the impulse $J = F\Delta t$. By the impulse-momentum equation, we now have $F\Delta t = 4600/g \approx 144$ lb-s, where F is the force that we want to compute. We only need a value for Δt. A reasonable guess is $\Delta t = .2$ s. Then $F * .2 \approx 143.75$ or $F \approx 144/.2 = 720$ lb.

Is 720 pounds more or less than you were expecting? On the one hand, 720 pounds worth of barbells would flatten most people in the gym. On the other hand, you may have seen an episode of *Sport Science* in which sensors measured a Ray Lewis hit on a crash test dummy at 2200 pounds. Factors that can increase the calculated force are the size and speed of the player, the impact time, and the velocity after the collision. In the *Sport Science* collision, the dummy was driven backwards by Lewis. The next example picks up the story at that point.

Example 4.2 A 160-pound crash dummy moving 14 mph is tackled by Ray Lewis, who delivers a force of 2200 pounds for 0.1 s. At what speed is the dummy moving right after the collision?
Solution. Convert the dummy's speed to ft/s, so that $v = 14(5280/3600) \approx 20.5$ ft/s and $m = 160/32 = 5$ slugs. The dummy's linear momentum before the collision is $p = 20.5(5) = 102.5$ lb-s. The impulse for the hit is $J = F\Delta t = 2200(.1) = 220$ lb-s. Since impulse equals the change in linear momentum, the dummy's linear momentum after the collision is $102.5 - 220 = -117.5$ lb-s. If the dummy's velocity after the collision is denoted by w, then $5w = -117.5$ or $w = -23.5$ ft/s, where the negative indicates that the dummy is moving in the opposite direction. Lewis did not merely stop the dummy's progress, he knocked it in the opposite direction at a faster speed than it had been moving forward! In this case, Lewis would need to wrap up the dummy with his arms to complete the tackle, or Lewis and the dummy would fly apart.

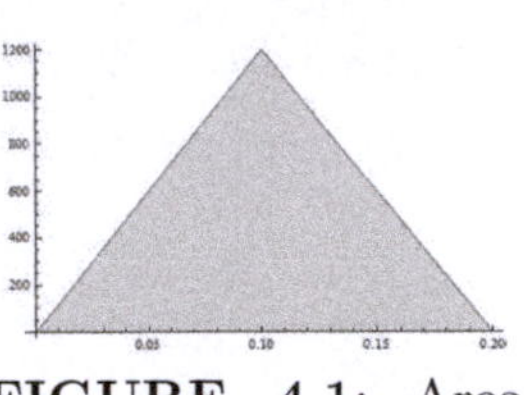

FIGURE 4.1: Area for Nonconstant Force

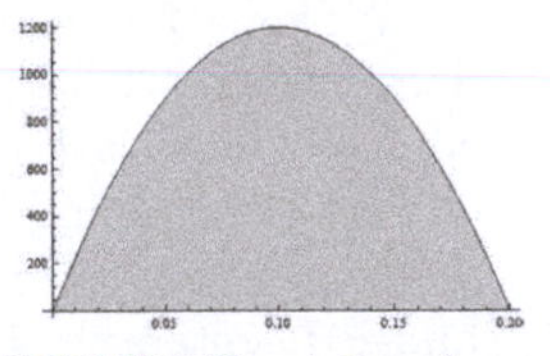
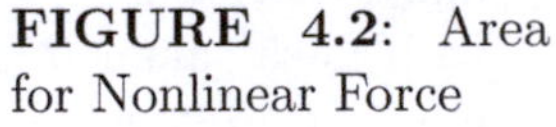

FIGURE 4.2: Area for Nonlinear Force

In the first two examples, we assumed that a constant force was applied for a period of time. In reality, the force builds up from zero to a maximum and then decreases back to zero. For a non-constant force, the calculation of impulse requires some geometry and, in general, calculus. However, the underlying concept is simple. Figure 4.1 shows a force that lasts for 0.2 s, building linearly from 0 lb to a maximum of 1200 lb at time 0.1, then dropping linearly to 0. The impulse equals the area "under the curve," in this case the shaded triangle formed by the graph and the horizontal x-axis. In Figure 4.2, we have the same basic situation of a force building up to a peak of 1200 pounds at time 0.1 s before dropping back to 0 pounds. This time, the buildup is non-linear (in fact, the graph is a parabola). We can argue in two different ways that the impulse for this force is larger than that of the force in Figure 4.1. First, it should be clear visually that the shaded area is larger in Figure 4.2 than in Figure 4.1. Second, for any given time value the graph of the parabola is above the straight line segments in Figure 4.1. This means that the force is larger, and therefore will create a larger impulse.

Exact values can be computed in many cases.

Example 4.3 Compute the impulse for the forces in Figure 4.1 and Figure 4.2, and for a constant force of 1200 pounds for 0.2 s.
Solution. The impulses equal the areas under the curves, as shaded in the figures. The area of a triangle is 1/2 base times height. In Figure 4.1, the height is 1200 and the base is 0.2, so the area is $.5(1200)(0.2) = 120$ and the impulse is 120 lb-s. In Figure 4.2, we have the area bounded by a parabola. This

calculation requires calculus (see the calculus box below) or **Archimedes' rule** that the area is 2/3 times base times height; in this case, the area is $\frac{2}{3}(1200)(0.2) = 160$ and the impulse is 160 lb. For a constant function, the graph forms a rectangle with area equal to base times height, or $1200(0.2) = 240$ with an impulse of 240 lb-s. Note in this last case that we can use the constant force formula $F\Delta t$ to get the impulse.

Before taking a quick look at the calculus of computing areas, we ask a question. How should the forces in Example 4.3 be reported? They all reach a peak of 1200 pounds, but their effects as measured by impulse are quite different. Therefore, saying that each is a 1200-pound force could be misleading. One resolution is to compute the average force in each case. Since a constant force of 600 lb applied for 0.2 s produces an impulse of 120 lb-s, we can say that the average force in Figure 4.1 is 600 lb. Similarly, the average force in Figure 4.2 is the impulse divided by the time span, or 160 lb-s / 0.2 s = 800 lb. This clearly indicates that this force has a bigger impact. However, if the point is to impress the audience with the largest number possible, the peak force of 1200 lb is more impressive than the average force of 800 lb.

Calculus Box: Integration

One of the most important techniques of one variable calculus is integration. In the case of a positive function $f(x)$, the (definite) integral from $x = a$ to $x = b$ gives the area under the graph of f, above the x-axis, and between $x = a$ and $x = b$. If we accept at face value that impulse is given by such an area, then we can conclude that impulse is computed with an integral. However, this avoids the obvious question of why impulse equals this area. It is helpful to derive the formula for impulse from first principles.

For a constant force F over a length of time Δt, impulse is given by $F\Delta t$. A force of 1200 lb for 2 s has an impulse of 1200(2)=2400 lb-s. Let's make the force non-constant by saying that for the first second the force is 800 lb and for the second second the force is 1000 lb. The impulses are 800 lb-s and 1000 lb-s, which add to a total impulse of 1800 lb-s. Now, suppose that we have different forces for each tenth of a second: F_1 lb from time 0 s to time 0.1 s, F_2 lb from time 0.1 s to time 0.2 s, and so on through F_{20} lb from time 1.9 s to time 2.0 s. Then the total impulse is

$$F_1(.1) + F_2(.1) + ... + F_{20}(.1) = \sum_{i=1}^{20} F_i(.1)$$

We now generalize to a force $F(t)$ on an interval $a \le t \le b$. Divide the interval into n subintervals of equal length $\Delta t = \frac{b-a}{n}$. If n is large, we expect that the force F is approximately constant on each subinterval, so that $F(t_i)\Delta t$ is a good approximation to the impulse on the i-th subinterval S_i, with t_i being a point in S_i. Then the impulse J is approximately

$$J \approx \sum_{i=1}^{n} F(t_i)\Delta t.$$

The larger n is, the better the approximation should be. In the limit,

$$J = \lim_{n\to\infty} \sum_{i=1}^{n} F(t_i)\Delta t = \int_a^b F(t)\,\mathrm{d}t$$

so that the integral comes into play because we approximate the impulse as the sum of impulses over small time intervals. In the limit, this sum approaches an integral, which (because force is a non-negative quantity) equals area under the curve.

Example 4.4 For the force $F(t) = 120000t(.2 - t)$, $0 \le t \le 0.2$, in Figure 4.2, compute the impulse.
Solution. By the above argument, the impulse is

$$\int_0^{0.2} 120000(.2t - t^2)\,\mathrm{d}t = 120000(.1t^2 - t^3/3)\Big|_{t=0}^{t=0.2}$$

$$= 120000\left[.1(.2)^2 - (.2)^3/3\right] = 160 \text{ lb-s}$$

as we previously obtained using Archimedes' rule.

Giving to Receive

The impulse-momentum equation can be used to explain some basic techniques in sports. If you have ever played catch, think for a moment about how you catch a baseball. First of all, you want to use a glove. Further, you probably have a nearly unconscious habit of pulling your hand back when the ball arrives. Both are good ideas, as we see below.

To catch the ball, you must remove all of its linear momentum. This change of linear momentum equals the impulse you apply to the ball (and, by Newton's third law, the same impulse is applied back to you). Any combination of force and time can accomplish this, as long as $F\Delta t$ equals the desired quantity. The important point is this: if you can increase Δt, the impact time, you can decrease the force and achieve the same impulse. A padded glove slows the ball down and lengthens the impact time. Pulling your hand back and giving with the ball also increases the contact time. In both cases, the longer contact time requires less force. Your hand appreciates the reduction in force.

A different aspect of the same principle occurs when a soccer player tries to trap a long pass. If the player is able to "catch" the ball with a foot that gives with the ball, the force applied to the ball is reduced. Then the change in linear momentum is reduced from a large change that sends the ball in the opposite direction to a smaller change that brings the ball down with zero velocity.

Tendons and Tennis

Impulse can be computed from the graph of force as a function of time. A different graph, also of interest in sports, shows force as a function of distance. Our feet endure constant compression and stretching as we move around. The ground pushes the balls of our feet, while our Achilles tendons pull on the backs of our feet. A specific force, shown on the vertical axis, causes a specific compression of the foot, shown on the horizontal axis. Figure 4.3 is based on data from Alexander's *Exploring Biomechanics*, where the foot compression is measured in millimeters and the force is measured in kilo-Newtons.

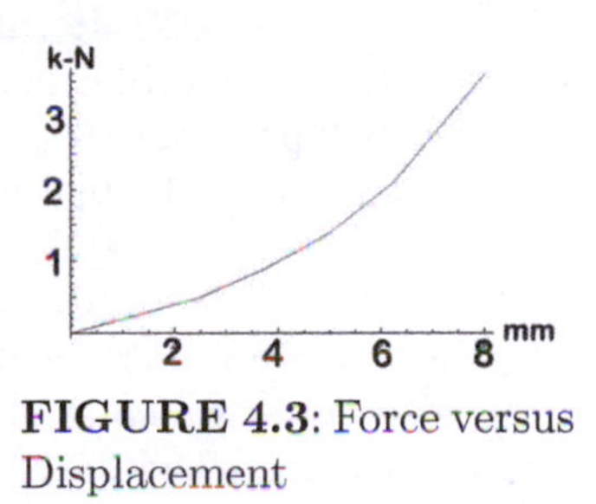

FIGURE 4.3: Force versus Displacement

After the foot is compressed by these forces, it springs back to normal length. In doing so, it pushes back against the ground and Achilles tendon. Alexander also measured these forces, which form the lower of the two curves in Figure 4.4. Notice that the forces during relaxation of the foot are smaller than the forces during compression. The reduction in force is a product of the loss of energy to heat in the foot. The area under each curve is proportional to the energy change. If A_c is the area under the compression curve and A_r is the area under the relaxation curve, then the ratio A_r/A_c is the proportion of energy retained in the compression/relaxation cycle. In other words, this is the efficiency of the foot at retaining energy. Alexander measured this value at about 78%. By comparison, tendons from the hind legs of wallabies retain about 93% of their energy, making wallabies better jumpers than humans.

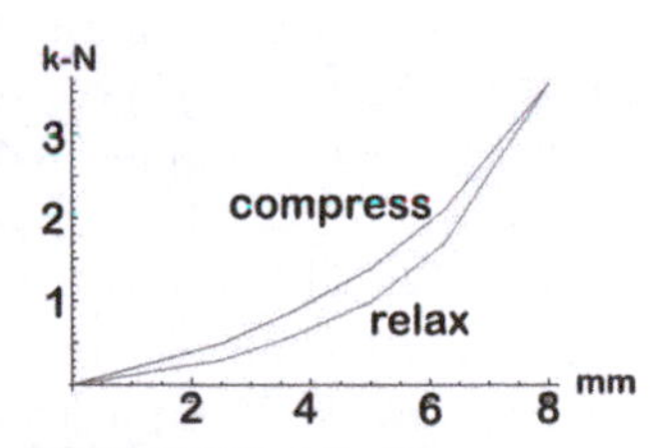

FIGURE 4.4: Force versus Displacement

This analysis also applies to collisions in which balls compress and relax.

When a tennis ball hits the strings of the tennis racket, the ball flattens almost completely before popping back out to its spherical shape. The same phenomenon happens, to a lesser extent, to golf balls and baseballs in their collisions with clubs and bats. The proportion of energy lost varies from situation to situation. For collisions of balls with other objects, we redefine the energy loss in a convenient way.

Coefficient of Restitution

Think of dropping a ball from some height. It falls to the floor and then bounces back up, but does not make it back to its original height. The next bounce is lower still. It turns out that there is a regularity to the bounces. If the ball reaches 70% of its original height on the first bounce, the second bounce will be about 70% of the first bounce (now 49% of the original height). On each bounce, the ball retains about 70% of its energy.

Recall that the height that a ball is dropped from determines its speed when it hits the ground, and the speed at which a ball is launched from the ground determines its height. Since the heights are different, it must be that the speed v_b at which the ball hits the ground is greater than the speed v_a at which it is launched back into the air. The ratio of the speeds is named the **coefficient of restitution (COR)**. That is,

$$\text{COR} = \frac{v_a}{v_b}.$$

The definition can apply to any collision of two objects, with v_b equalling the *relative velocity* before the collision and v_a the relative velocity after the collision. For example, if right before the collision a bat is moving 80 mph in one direction (the negative direction) and a ball is moving 90 mph in the opposite (positive) direction, then $v_b = 90 - (-80) = 170$ mph. If the bat and ball exit the collision moving in the same direction with ball speed 120 mph and bat speed 35 mph, then $v_a = 120 - 35 = 85$ mph, and the COR for this collision is $\frac{85}{170} = 0.5$.

Notice the wording of the last statement. It is important to realize that COR is a property of *the collision of two objects* and not just one object. Therefore, it does not make sense to ask for the COR of a baseball. You will get a higher bounce (higher COR) bouncing a baseball off wood than off a pillow. For brevity, we might say that the COR of a baseball is 0.546, but this is only valid for certain types of collisions. (In the case of major league baseballs, COR is measured for balls bouncing off of wood at 85 ft/s.) We sometimes assume, given no other data, that the COR for other types of collisions is approximately the same. However, the COR depends on many factors, including the speeds of the objects involved and their composition.

From the definition, it should be clear that COR is a number between 0 and 1. If COR = 0, then the collision is called perfectly inelastic and the objects stick together; an example is a football tackle. If COR = 1, then the collision is called elastic and no energy is lost. Our interest will be in inelastic collisions with $0 < \text{COR} < 1$. In general, the larger the COR is the bouncier or livelier the collision is. This has important ramifications in almost all ball sports.

Incoming and Outgoing

The situation to be analyzed in this section applies to many sports. To make the discussion concrete, imagine a baseball being hit by a bat. Before the collision, the ball and bat move in opposite directions with speeds v_{ball} and v_{bat}, respectively. After the collision, the ball and bat move in the same direction with speeds w_{ball} and w_{bat}, respectively. The situation is depicted in Figure 4.5.

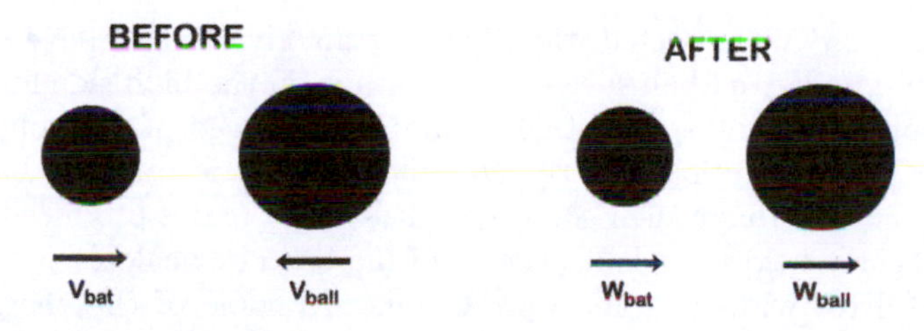

FIGURE 4.5: Speeds Before and After

Two principles and some algebra give us some insight into the interactions of ball and bat. The first is COR, which in this context is given by

$$\text{COR} = \frac{w_{ball} - w_{bat}}{v_{bat} - -v_{ball}} \tag{4.1}$$

where the double negative in the denominator is due to the ball moving in the negative direction. The second principle is conservation of linear momentum, which in this case means

$$m_{bat}\, v_{bat} - m_{ball}\, v_{ball} = m_{bat}\, w_{bat} + m_{ball}\, w_{ball} \tag{4.2}$$

where we again account for the direction of the moving ball.

Typically, we want to know how the ball speed after the collision depends on the other factors. For example, how can we give the ball enough speed to clear the fence for a home run? We can solve equation (4.1) for w_{bat} and substitute into equation (4.2). After some rearrangement, we get

$$w_{ball} = \frac{m_{bat}(COR\, v_{ball} + (1 + COR)\, v_{bat}) - m_{ball}\, v_{ball}}{m_{bat} + m_{ball}}. \quad (4.3)$$

This looks complicated, but it is actually easy to use and interpret.

Example 4.5 For a baseball of weight 5.25 oz and speed 90 mph, a baseball bat of weight 32 oz and speed 80 mph, and a COR of 0.4, find the speed of the ball off the bat.
Solution. From equation (4.3), the speed is

$$w_{ball} = \frac{32(0.4 \cdot 90 + 1.4 \cdot 80) - 5.25 \cdot 90}{32 + 5.25} = 114.5 \text{ mph}.$$

Explanations are in order. We did not convert weight to mass, or speeds to ft/s. An examination of the units shows why. COR is unitless, so $w_{ball} = \frac{32 \text{ oz} (0.4 \cdot 90 \text{ mph} + 1.4 \cdot 80 \text{ mph}) - 5.25 \text{ oz} \cdot 90 \text{ mph}}{(32+5.25) \text{ oz}}$. The "oz" units cancel, as any unit of mass would have. That leaves units of "mph." This is nice! As long as we're consistent, the units do not matter.

Easier than you expected, right? Unfortunately, we still have some issues to resolve. The batted ball speed of 114 mph is on the high side for the major leagues, but the pitch speed of 90 mph and bat speed of 80 mph are on the low side. An assumption underlying equation (4.3) is that the ball and bat collide in a line through their centers of mass. A fly ball is produced by the bat hitting slightly below the ball's center of mass, so our calculation only holds for a line drive with no spin. More commonly, some of the energy transfer from the bat to the ball is diverted from ball speed to ball spin. Further, the major league specification for a baseball is that the COR must be between 0.514 and 0.578. The value of 0.4 used in Example 4.5 is too low. I chose it, to be honest, so that the calculated ball speed was low enough to be plausible. To justify this change, note that the major league testing procedure is to fire a ball at 85 ft/s at a wooden wall. The relative speed of ball and bat before the collision in Example 4.5 is 170 mph, about 250 ft/s. The large difference in laboratory and playing field velocities makes it likely that the official COR of about 0.55 is too large.

In Example 4.5, we used a bat weight of 32 oz. This is close to the average weight in the major leagues, but think about the physical situation. The bat doesn't swing itself at 80 mph. A person is attached to the bat. Shouldn't we include the mass of the batter as well as the bat? The answer is that the m_{bat} term in equation (4.3) is actually a new quantity called **effective mass**. The effective mass of the bat depends on where on the bat the ball is hit. If it is hit in line with the center of mass of the bat, then effective mass and mass of the bat are the same. Generally, on off-center hits, the effective mass of the bat is less than the mass of the bat.

Brody, Cross, and Lindsey introduce detailed information about tennis rackets in *The Physics and Technology of Tennis.* In the racket-ball collision, the effective mass issue is especially confusing, since the strings are the only part of the racket that touch the ball. The center of mass of the racket is near the neck, barely on the strings. The racket's maximum COR occurs at this spot, and effective mass equals racket mass here. Moving up the racket reduces effective mass. By the time you reach the center of the racket, the effective mass is reduced to about half the mass of the racket.

An interesting experiment illustrates the role of grip on effective mass. Stand a tennis racket on its end and throw a ball at the center of the strings. The racket will be knocked over, but mark where the ball first bounces. Then repeat the process with someone holding the racket tightly by its handle. The ball will bounce to the same spot! The ball speed is not affected at all by the complete absence of a grip! The effective mass does not depend on human interaction: the strings stretch and then send the ball on its way before the vibrations of the racket can reach and return from the hand. The ball has left before the strings know whether or not your hand is there.

It has been shown that the effective mass for a football field goal kicker is larger if the kicker kicks from the side, soccer-style, rather than head on. This is one reason that all kickers, even those who never played soccer, kick from the side.

Derivative Works

What effect does a decrease in effective mass have? These and other basic questions can be answered with some algebra and/or calculus. For example, take equation (4.3) and think about the effect of increasing m_{bat}. Since m_{bat} multiplies a positive constant in the numerator, the numerator will increase. But so will the denominator. We need calculus here, in particular the derivative. As is shown in the calculus-based exercises, an increase in the effective mass of the bat or racket always results in an increase in ball speed. This should coincide with your experience. Swing a big stick to hit it hard!

It is easier, mathematically, to see what happens if v_{bat} is increased. Since v_{bat} is found only in the numerator multiplied by a positive constant, an increase in v_{bat} results in an increased ball speed. This should make sense: a faster swing has the potential for the hitting the ball harder (if you can make contact). An interesting analysis involves changes to v_{ball}. Since v_{ball} is multiplied by $m_{bat}\,\mathrm{COR} - m_{ball}$, an increase in v_{ball} increases the ball speed if $m_{bat}\,\mathrm{COR} - m_{ball} > 0$, but decreases the ball speed if $m_{bat}\,\mathrm{COR} - m_{ball} < 0$. In baseball and most sports, $m_{bat}\,\mathrm{COR} - m_{ball} > 0$, so the faster the ball comes in, the faster it goes back out.

The Way the Ball Bounces

One of the most common and important collisions in sports is one between the ball and ground. The mechanics here are more complicated than in equation (4.3), because the ball typically moves at an angle to the center of mass of the striking object (Earth). However, we can determine some basic principles from Newton's laws.

In the figure, a ball hits the ground while moving to the right and downward with no spin. When it hits the ground, the ball pushes to the right and downward on the ground, which (by Newton's Third Law) pushes the ball upward and to the left. The leftward push causes the ball to spin in a direction that we recognize as topspin.

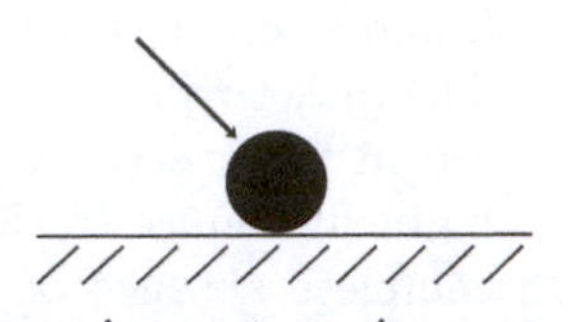

The ball leaves the ground with topspin, at an angle that is different from the incoming angle. The overall speed of the ball has been reduced, but the upward and leftward push of the ground gives the ball relatively more vertical motion and less horizontal motion; the angle to the ground will be greater outgoing than incoming. The amounts that speed and angle change are determined by the surface properties of ball and ground. The greater the coefficient of friction, the longer the ball and ground stay in contact and the more effect the ground has.

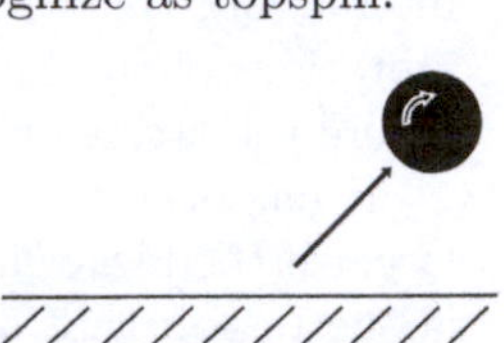

If the incoming ball has backspin, the bottom of the ball in the above figure is moving left-to-right. The backspin ball then pushes the ground harder to the right than the no-spin ball, and receives a harder push to the left in return. If the pushback to the left is hard enough, the ball can bounce backwards. You have probably seen this effect with a tennis ball, ping pong ball, baseball, or golf ball. Until 2010, you rarely saw this in football, but punters have now started putting backspin on the ball (by kicking it with the nose down) to keep punts from bouncing into the endzone.

Backspin can explain the phenomenal success of Rick Barry shooting free throws. One of the all-time greats in the NBA and ABA, in his last eight seasons Barry made 2496 out of 2731 free throws (91.4%!) while leading the league in percentage made in six of the eight years. Find a video of Barry shooting free throws: he used an underhand motion! One of the advantages of this technique is the ability to put extra backspin on the ball. If the ball hits the rim, the backspin will result in a softer bounce that increases the chances that the ball goes in.

One aspect of the backspin bounce may seem counterintuitive to tennis players. The stronger leftward push of the ground takes more horizontal velocity away from a backspin ball than a no-spin ball, so that the angle of the bounce is *greater* than that of the no-spin ball. Tennis players hit slices with lots of backspin to keep the ball low. How does this work if backspin makes the ball bounce higher? The above graphics have confused the issue. Because backspin creates an upward Magnus force, the trajectory of the ball can be much flatter than a no-spin shot, so that its incoming angle can be much smaller than that of a typical shot. The small incoming angle is what keeps a backspin slice low, even though it bounces at a higher angle than a no-spin shot with the same incoming angle.

Tennis is played on many different types of surfaces: grass and clay are the old traditionals, but a variety of synthetic courts allow tournaments to control the pace of play. How can we measure the pace of a tennis court? You might start with COR; if the ball loses most of its speed as the result of a small COR, then the court must be slow, right? Unfortunately, grass courts have much smaller COR's than do clay courts. (Think about it: would you expect a ball to bounce higher off of grass or clay?) This contradicts our knowledge that grass is fast and clay is slow. Another factor that influences court speed is the coefficient of friction. This measures the resistance to motion while the ball is in contact with the ground. Grass is slick with a low coefficient of friction, while clay is rough with a high coefficient of friction. A quantity named *Court Pace Rating* combines the two coefficients (COR and friction) into a meaningful rating.

Freeze Frame

One of the most interesting revelations of sports high-speed photography is the choreography of a hockey slap shot. The player takes a big windup and slaps the stick into the ice right behind the puck, with one hand positioned near the bottom of the stick. The low positioning of the bottom hand reduces the velocity of the stick at impact, but the power of the slap shot comes more from potential energy than kinetic energy. Notice the extreme curve of the stick from top hand to bottom hand in the picture.

The bottom hand pressing down on the stick causes it to bend substantially, storing potential energy. As the stick flexes back into shape, the puck is swept along. In the photograph to the right, notice how far along the stick and puck are without separating. The stick's potential energy is transferred to the puck as kinetic energy. A *Sport Science*

episode measured a shooter's arm speed at 75 mph, stick speed at 80 mph, and puck speed at 100 mph. Photographs have shown the stick and puck making contact multiple times. One problem with an extreme bending of the stick is that too much bending will cause the sticks to break. Phil Kessel of the Maple Leafs and other players required specially made sticks with extra strength. Modern sticks are made with flexible lightweight carbon materials and are designed for optimal flex. Manufacturers proclaim that their sticks will "load up" on slap shots, consistent with our analysis of the stick-puck collision.

Exercises

In these exercises, Ⓣ refers to thinking problems, conceptual problems requiring no calculations. Ⓒ refers to problems requiring calculus or significant computer calculations. Ⓟ refers to projects; these are ideas for further investigation (hints and resources are at the book's web site).

4.1 A 220-pound running back moving 18 mph runs into a 320-pound defensive lineman. (a) What is the speed needed for the lineman to stop the running back? (b) Assuming that the collision lasts 0.2 second, what is the average force? (c) Find the peak force if the force is piecewise linear as in Figure 4.1. (d) Find the peak force if the force is parabolic as in Figure 4.2.

4.2 A 180-pound defensive back moving 18 mph runs into a 240-pound running back moving at 12 mph. (a) What is the outcome of the collision? (b) Assuming no other forces, what is the combined velocity of the pair after the collision? (c) Assuming that the collision lasts 0.2 second, what is the average force? (d) Find the peak force if the force is piecewise linear as in Figure 4.1. (e) Find the peak force if the force is parabolic as in Figure 4.2.

4.3 Find the impulse for each force. (F is in pounds, and t is in seconds.)
(a) $F(t) = 100$ for $0 < t < .1$; $F(t) = 300$ for $.1 \leq t \leq .2$; $F(t) = 600$ for $.2 \leq t \leq .4$; $F(t) = 300$ for $.4 \leq t \leq .5$; $F(t) = 100$ for $.5 \leq t \leq .6$;
(b) $F(t) = 200t$ for $0 < t < 2$; $F(t) = 1200 - 400t$ for $2 < t < 3$
(c) $F(t) = 400t(3 - t)$ for $0 < t < 3$
(d) $F(t) = 600t(2 - t)$ for $0 < t < 2$

4.4 Given the following compression/relaxation profiles for tendons (see Figure 4.4), which one represents the more efficient tendon? If the curves instead are for two tennis balls, which ball is livelier?
(a) $\frac{1}{2}(x^2 + x)$ and x^2 for $0 < x < 1$ (b) $\frac{1}{2}(x^2 + x)$ and x^4 for $0 < x < 1$

4.5 The following statements give official regulations for sports equipment. For each, give the speed of impact, how the speed at impact compares to typical collision speeds in that sport, and the range of CORs.
(a) A basketball dropped from a height of 1.8 m bounces to a height between 1.2 m and 1.4 m.

(b) A lacrosse ball dropped from a height of 72 in bounces to a height between 43 in and 51 in.
(c) A tennis ball dropped from a height of 100 in bounces to a height between 53 in and 58 in.
(d) A high-altitude tennis ball dropped from a height of 100 in bounces to a height between 48 in and 53 in.

4.6 To control the "trampoline effect" of hollow metal drivers, golf now legislates the COR of a driver-ball collision to be no more that 0.83. If a driver is moving at 140 mph, find the maximum relative speed of the ball and driver after the collision. Is this the same as the maximum ball speed?

4.7 Show that COR $= \sqrt{\frac{a}{b}}$ where a is the height of the bounce of a ball dropped from height b.

4.8 (a) For the ball and bat of Example 4.5, find the ball speed w_{ball} if the bat speed is increased to 81 mph.
(b) For the ball and bat of Example 4.5, find the ball speed w_{ball} if the incoming ball speed is increased to 91 mph.
(c) Which has the larger impact on ball speed, bat speed or ball speed?

4.9 (a) A 57 gm tennis ball moving 60 mph is hit by a 300 gm racket moving 80 mph in the opposite direction. If COR $= 0.74$, find the ball speed after impact.
(b) A golf ball of mass 0.05 kg at rest is hit by a golf club of mass 0.17 kg moving 120 mph. If COR $= 0.8$, find the ball speed after impact.

4.10 Solve equations (4.1) and (4.2) for w_{bat}.

4.11 Use the solution of exercise 4.10 to determine whether w_{bat} increases or decreases when (a) v_{ball} increases; (b) v_{bat} increases.

4.12 Find w_{bat} in exercise 4.9, parts a and b.

4.13 (T) Use the concept of impulse to explain good techniques for each:
(a) catching a football; (b) landing from a large height;
(c) dribbling a basketball; (d) hitting a drop volley in tennis.

4.14 (T) Use the concept of impulse to explain why running in sand is more difficult than running on concrete.

4.15 (T) Tennis rackets can be strung at different tensions. Given that looser strings create longer impact times, discuss whether looser or tighter strings would produce (a) more ball speed; (b) more control.

4.16 (T) Discuss what would happen to the effective mass of a field goal kicker if his planting foot slipped.

4.17 (T) The effective mass of a tennis racket does depend on the tightness of the player's grip if the ball is hit off-center (not on the line running through the handle and out the top of the racket). Explain.

4.18 (T) It was noted in the text that w_{ball} decreases with an increase in v_{ball} if $m_{ball} > m_{bat} \cdot$COR. Explain in physical terms why this makes sense.

4.19 (T) Two balls with positive horizontal (and negative vertical) velocity hit the ground at the same angle. Ball A has no spin, and ball B has topspin. Which ball bounces at a larger angle to the ground? Explain.

4.20 (T) Is it possible for a ball with topspin to have a larger horizontal velocity after hitting the ground than before? Explain. Discuss how this result applies to ground balls in baseball and ground strokes in tennis.

4.21 Ⓣ A ball moves horizontally with backspin before hitting a wall. Does the spin cause the ball to rebound off the wall higher or lower than a ball with no spin? What type of spin does the ball have after hitting the wall? Discuss how this result applies to shooting a basketball off the backboard.

4.22 Ⓣ Repeat exercise 4.21 with topspin. Discuss how the result applies to a tennis volley of a topspin shot.

4.23 Ⓣ A baseball player hits a fly ball by making contact with the top portion of the bat barrel hitting the bottom part of the ball. What type of spin is produced? Does this spin increase or decrease the distance of the fly ball?

4.24 Ⓣ Typically, a fastball has more speed than a curveball. Only taking this into account, explain why it should be easier to hit a home run off of a fastball. For which pitch would a fly ball have more backspin? Simulations have shown that a well-struck hit off of a curveball will travel farther than off of a fastball. Briefly explain how this could be true.

4.25 Ⓣ A hockey slap shot starts with the player taking a big swing and hitting the ice right behind the puck. Explain why a similar technique on the golf course would not be effective. Use spin to explain why golfers get better results hitting the ball directly.

4.26 Ⓣ Compared to grass and clay, describe how large the COR of a basketball court would be for a tennis ball. How large would its coefficient of friction be? How large would its Court Pace Rating be?

4.27 Ⓣ In the *Sport Science* episode "Human Flight" Jerry Rice talks about catching footballs with the fingertips: "If it hits you in the palms of your hands you're going to have that ricochet." Discuss the advantage of fingertips over palms in terms of impulse.

4.28 Ⓒ Compute the impulse and average force of each force. (a) $1,000,000t^2(0.2-t)$ for $0 < t < 0.2$. (b) $1000\sin(\pi t/0.2)$ for $0 < t < 0.2$.

4.29 Ⓒ Prove Archimedes' Rule for $F(t) = at(b-t)$ for $0 < t < b$ with $a > 0$ and $b > 0$.

4.30 Ⓒ For a constant force F applied over a distance d, *work* is defined by $W = Fd$. A force $F(x)$ is applied at location x for $a \le x \le b$. Derive the more general formula $W = \int_a^b F(x)dx$.

4.31 Ⓒ Use the data to estimate the efficiency $1 - \dfrac{\int_a^b f_c(x) - f_e(x)dx}{\int_a^b f_c(x)dx}$.

(a) tennis ball data

x in	0.0	0.1	0.2	0.3	0.4
f_c lb	0	25	50	90	160
f_e lb	0	23	46	78	160

(b) baseball data

x in	0.0	0.1	0.2	0.3	0.4
f_c lb	0	250	600	1200	1750
f_e lb	0	230	450	700	1750

(c) Wallaby tendon data

x mm	0.0	0.75	1.5	2.25	3.0
f_c N	0	110	250	450	700
f_e N	0	100	230	410	700

(d) human foot data

x mm	0.0	2.0	4.0	6.0	8.0
f_c N	0	300	1000	1800	3500
f_e N	0	150	700	1500	3500

4.32 Ⓒ Use equation (4.3) to compute the derivative of w_{ball} with respect to m_{bat}. That is, treat m_{bat} as the variable and all other parameters as constants. Show that the derivative is positive, and interpret this to mean that a bigger bat will hit a ball harder.

4.33 Ⓒ Compute the derivative of w_{ball} with respect to m_{ball}, determine if it is positive or negative, and interpret the result.

4.34 Ⓒ Use the solution to exercise 4.10 to compute and interpret the derivative of w_{bat} with respect to (a) m_{bat}; (b) m_{ball}

4.35 Ⓒ In Example 4.5, $w_{ball} \approx 0.2v_{ball} + 1.2v_{bat}$. Explain why the increase in w_{ball} in exercise 8(a) is 1.2 mph and the increase in exercise 8(b) is 0.2 mph.

4.36 Ⓒ In Example 4.5, $w_{ball} = \dfrac{148m_{bat} - 90m_{ball}}{m_{bat} + m_{ball}}$.
(a) Use this formula to compute the derivative of w_{ball} with respect to m_{bat} and interpret the result in terms of the change in w_{ball} for a change in m_{bat}.
(b) Repeat part (a) with respect to m_{ball}.

4.37 Ⓟ Drop a ball from different heights and construct a graph of COR as a function of impact velocity. Does the graph appear to be linear?

4.38 Ⓟ Bounce balls off of different places on a tennis racket to determine COR as a function of location on the racket.

Further Reading

Peter Brancazio's *Sport Science* is an excellent introduction to the physics of sports. The bits on catching a ball, throwing a ball at a tennis racket, and the effective mass of a field goal kicker are from Brancazio.

Robert Watts and Terry Bahill's *Keep Your Eye on the Ball* is the best of several physics of baseball books.

A search for "Physics of (favorite sport)" will bring up many results. I can recommend *The Physics of Basketball* (Fontanella), *The Physics of Hockey* (Hache), *Football Physics* (Gay), *Newton's Football* (St. John and Ramirez), *The Science of Soccer* (Wesson), *The Physics and Technology of Tennis* (Brody, Cross, and Lindsey), *Golf Science* (Smith), *The Science of Golf* (Wesson), *Golf By the Numbers* (Minton), and *The Physics of NASCAR* (Leslie-Pelecky).

Football Physics is the source of impact time for a football collision.

Several episodes of John Brenkus's *Sport Science* are available at http://espn.go.com/espn/sportscience/. Accessed 8-11-2015. The Ray Lewis "Block and Tackle" segment was referenced in the text. All segments are informative and entertaining.

I am, of course, partial to Smith and Minton's *Calculus* as a calculus reference. It has numerous sports problems.

R. McNeill Alexander's *Exploring Biomechanics* is the source for the information about wallaby tendons and human feet.

Before co-authoring *The Physics and Technology of Tennis*, Howard Brody published *Tennis Science for Tennis Players.*

Austin Murphy's article "The Nail in the Coffin" about Aussie-style punting with backspin appeared in the October 27, 2014 issue of *Sports Illustrated.*

Lee Torrey's *Stretching the Limits* has a nice description of the hockey slap shot. Several excellent bent-stick pictures are online. A search for "physics of hockey slap shot" will turn up good sites such as http://physicsofhockeyproject.weebly.com/shooting.html. Accessed 8-11-2015.

A good article from the *Wall Street Journal* about hitting curveballs farther than fastballs is at http://www.wsj.com/articles/SB108568923118723199. Accessed 8-11-2015.

Chapter 5

Ratings Systems

Introduction

College Football Ranking Composite

Saturday, November 9, 2013 (124 Rankings)

| Massey Ratings | Last Week | Archived | FBS | FCS | BCS | CSV Data | About |
Meta Rank Analysis: | Kislanko | Wobus | Pollstalker | Rankings | Superlist |

Payne
OSCAR
Coffey
Minton
RoundTable

AccuRatings
Laz Index
Round Robin Win %
Rudacille
Ashburn

Adjusted Stats EWP
ARGH
GRS Gindin
Sport Theory
Kellner MOV

Self
Kirkpatrick
Massey
Burrus
Breisch

PAY	OSC	COF	MTN	RT	ACU	LAZ	RWP	RUD	ASH	Rank, Team,	Conf,	Record	HKB	ARG	GRS	STH	KMV	SEL	KPK	MAS	BU
1	1	1	1	1	2	1	2	1	2	1 Florida St	ACC	9-0	1	2	1	1	2	2	1	2	
3	3	3	3	3	1	2	1	2	1	2 Alabama	SEC	9-0	3	1	2	3	1	1	3	1	
2	2	2	2	2	5	5	3	3	5	3 Baylor	B12	8-0	2	4	4	2	7	3	2	3	
5	6	6	4	5	3	4	7	5	4	4 Stanford	P12	8-1	6	6	5	5	3	5	5	5	
4	5	4	6	4	7	7	4	6	6	5 Ohio St	B10	9-0	5	3	7	6	4	6	9	7	
6	4	5	5	6	4	3	5	4	3	6 Oregon	P12	8-1	4	5	3	4	5	4	4	4	
7	7	7	8	7	6	6	6	7	8	7 Missouri	SEC	9-1	7	7	6	7	6	8	6	6	
10	9	10	9	9	10	11	8	11	10	8 Clemson	ACC	8-1	9	10	8	11	8	10	8	14	1
9	10	9	10	10	12	9	11	9	9	9 Auburn	SEC	9-1	11	8	9	10	9	9	11	9	
8	8	8	7	8	8	8	10	10	7	10 Arizona St	P12	7-2	8	9	10	8	13	7	7	10	
PAY	OSC	COF	MTN	RT	ACU	LAZ	RWP	RUD	ASH	Rank, Team,	Conf,	Record	HKB	ARG	GRS	STH	KMV	SEL	KPK	MAS	BU
11	12	12	12	11	15	10	13	12	12	11 Texas A&M	SEC	8-2	12	23	11	12	15	12	15	11	1
12	11	11	14	12	13	12	9	8	14	12 Wisconsin	B10	7-2	10	11	14	9	19	11	18	8	1
13	14	14	11	13	11	14	15	17	13	13 South Carolina	SEC	7-2	15	13	12	14	10	13	10	15	1
16	13	13	15	19	9	13	14	15	15	14 Oklahoma St	B12	8-1	14	14	21	18	11	24	14	16	1
22	18	20	17	18	16	17	12	21	16	15 Michigan St	B10	8-1	27	12	30	20	14	19	20	21	1
15	15	16	13	20	17	15	19	13	11	16 UCLA	P12	7-2	13	16	13	13	16	14	19	12	1

From 1998 to 2013, the participants in college football's national championship game were partially determined by computer ranking systems. The use of computers by the Bowl Championship Series (BCS) system was often ridiculed by football reporters. However, there was an explosion of ranking systems during the BCS era. The above graphic is a screen capture from the excellent Massey Ratings web site, showing some of the 124 ranking systems submitted during the week of November 9, 2013. Presumably, this is just the tip of the rankings iceberg.

In this chapter, we develop several matrix-based systems for rating sports teams or individuals. We will address a number of questions. What is the goal of a ranking system? What is the difference between a rating and a ranking? Are computer systems complicated? How is strength of schedule incorporated into the systems? How can ranking systems be evaluated?

Right versus Best

Imagine that you are in charge of ranking teams in your favorite sport. Are you more interested in picking the "right" teams or the "best" teams? The distinction may be unclear at first, but consider the following two teams. Team A has the best record in the league. However, its schedule was easy and it often came from behind to post narrow victories. Team B has the fifth best record in the league, against the toughest schedule in the league. Its losses were all early in the season with several players injured. With everybody healthy, team B finished the season winning its last several games by wide margins. The records make A the "right" number one, but the hot streak at the end makes B the "best" team.

Most leagues design ranking systems that emphasize the choice of the *right* team. Won-lost records dominate, and tie-breakers emphasize head-to-head matchups. This makes sense when the teams' schedules are equal or nearly equal. College sports teams can have wildly different strengths of schedule. Few people would choose an undefeated team from a minor league over a once-defeated team from a major league. Yet college football rankings often follow a "check-out line" mentality. If you are first in line, nobody is allowed to move ahead of you unless you lose. If you lose, the teams behind you all move up one slot in an orderly fashion. This type of ranking choreography largely ignores the issue of which team is actually the best.

A slightly different way of phrasing the question is to ask whether you are trying to evaluate the past or predict the future. Take teams A and B from above. If I seed team A #1 and team B #2 and then predict team B to win the game, it is clear that my ranking system is more about picking the right team based on past records than picking the best team based on who will win.

Here is the point: a mathematically sound analysis starts with clear objectives and finishes with a precise evaluation of how well the objectives are met. A major distinction between computer ranking systems and human ranking systems is that we humans rarely have a single set of objectives. We somewhat randomly balance the right choices with the best choices, sometimes using our predictions of the future but often not. An understanding of our basic irrationality is important when evaluating computer systems.

Ratings versus Rankings

The terms "rating" and "ranking" have thus far appeared in what may seem to be an interchangeable fashion. However, they are different. A **rating system** assigns a number to each team that corresponds to its strength. A

ranking system lists all of the teams in an order that corresponds to their strengths. A rating system can easily be turned into a ranking system by listing the teams in numerical order, but a rating system may provide extra information, such as how closely matched the teams are.

We assume that the ratings and rankings are **transitive**. That is, if we claim that A is better than B and B is better than C, then we implicitly acknowledge that A is better than C. This may seem obvious, but we will see in Chapter 6 that the issue is complicated. For now, note that our rating systems are one-dimensional in their output, producing a single number that characterizes each team's strength. We will not, for example, incorporate styles of play that could lead to a conclusion that while A will beat B and B will beat C, C matches up well with A and would beat A.

We will introduce three rating systems, two of which were part of the BCS computer ranking system. To get an idea of what a "computer system" is like, here are the main assumptions for each of the systems.

Massey System: a team's rating is the average of its opponents' ratings plus its net points per game.

Colley System: a team with w wins in g games has a rating of $\dfrac{1+w}{2+g}$, with w modified for strength of schedule.

Elo System: a team's rating is updated after each game, increasing after a win and decreasing after a loss, with the size of the change depending on the quality of the opponent.

The only intimidating aspect of any of these systems is the amount of time it would take to calculate the ratings by hand. That is why they are computer systems: computers are needed to do a sizable amount of number crunching.

The Massey System

The Massey System is named for Kenneth Massey, who developed it as an undergraduate student. Massey followed in the footsteps of others. The version presented here was published as a UMAP Module, and is a slight modification of work by Jech and others. It is interesting that this is the system that bears Massey's name, since in graduate school Massey developed a more sophisticated rating system that was part of the BCS system, and which he now applies to a variety of sports at the Massey Ratings web site.

Objective: The difference in ratings of two teams represents the point differential in a contest between the two teams.

This is a rating system, so each team is assigned a number representing its strength. The unit of measurement is points, and the model is based on the idea that if Dallas is rated 100 and Washington 88, then Dallas is 12 points better. The rating could be used to predict the outcome of a game between Dallas and Washington, or to rank Dallas above Washington.

Model: A team's rating is the average of its opponents' ratings plus its net points per game.

Thus, if Dallas has outscored its opponents by a total of 80 points in 16 games (an average of 5 points per game), its rating is the average of its opponents' ratings plus 5. To see how this plays out, we work this simple example.

Early in the season, a four-team league has the results shown to the side. "W" is wins, "L" is losses, "PF" is points for, and "PA" is points against. At first glance, you would think that A is the best team, D is the worst, and teams B and C are equal. This is not true, as we will see after getting some valuable information. What is missing?

Team	W	L	PF	PA
A	5	0	72	50
B	2	2	44	44
C	2	2	40	40
D	0	5	40	62

TABLE 5.1: Four-Team League

We need to know which games have been played, so that we can take into account strength of schedule. Take the schedule shown to the side. This tells us that A and B have played twice, A and C have not played, A and D have played 3 times, and so on. Now, why should we rank B above C?

	A	B	C	D
A	-	2	0	3
B	2	-	2	0
C	0	2	-	2
D	3	0	2	-

TABLE 5.2: Games Played

Team B played the best team in the league, while C did not. B had a tougher schedule, so its 2-2 record is more impressive than C's 2-2 record against inferior competition (two games against D, which B did not play). If you're not convinced, think through the results and note that since B lost twice to A, it must be that B won both games played against C!

The model for the Massey system translates to one equation for each team. Name the team ratings a, b, c, and d. Team A scored 22 points more than its opponents in 5 games, so team A's net points per game is $22/5 = 4.4$. Then a equals the average of A's opponents' ratings plus 4.4: $a = \frac{b+b+d+d+d}{5} + 4.4$. Multiply by 5 to get $5a = 2b + 3d + 22$ or $5a - 2b - 3d = 22$. Team B's net points per game is $0/4 = 0$ so its equation is $b = \frac{a+a+c+c}{4}$ or $4b - 2a - 2c = 0$. In alphabetical order, the complete set of equations is

$$5a - 2b - 0c - 3d = 22$$

$$-2a + 4b - 2c - 0d = 0$$

$$-0a - 2b + 4c - 2d = 0$$

$$-3a - 0b - 2c + 5d = -22$$

Notice how the schedule and results are encoded in the equations. For a given team's equation, its (unknown) rating is multiplied by its number of games played. Subtracted from that are the other teams' ratings, multiplied by the number of games played against that team. On the right-hand side of the equation is the net points scored (total, not per game). Easy!

All that's left is to solve the equations. We do so by *row reducing* the corresponding matrix. This is covered in a linear algebra course and (often) in precalculus. The matrix corresponding to the above equations is shown to the right.

$$\begin{bmatrix} 5 & -2 & 0 & -3 & 22 \\ -2 & 4 & -2 & 0 & 0 \\ 0 & -2 & 4 & -2 & 0 \\ -3 & 0 & -2 & 5 & -22 \end{bmatrix}$$

Each row of the matrix corresponds to one team's equation, and the entries in the rows are the coefficients of a, b, c, and d (in order!), followed by the number on the right-hand side. Notice again that the number of games against each team and the net points are readily apparent. Also, notice that each column sums to 0. This will cause us some grief shortly.

The reduced matrix is shown to the right.

$$\begin{bmatrix} 1 & 0 & 0 & -1 & 6 \\ 0 & 1 & 0 & -1 & 4 \\ 0 & 0 & 1 & -1 & 2 \\ 0 & 0 & 0 & 0 & 0 \end{bmatrix}$$

Each row can be translated back into equation form. The top row translates to $1a + 0b + 0c - 1d = 6$ or simply $a - d = 6$. Similarly, we have $b - d = 4$, $c - d = 2$ and $0 = 0$. At one level, that is hugely disappointing: we do not have a unique solution for the ratings. (This is a consequence of the original columns summing to 0, which means that the rows are linearly dependent.) Pick a value for d (*any* value) and you can find corresponding values for a, b, and c.

One way to obtain a unique solution is to add an additional requirement, such as the ratings summing to zero. However, our original objective was to find the *difference* between any two teams' ratings. The first equation tells us that A is 6 points better than D. Subtracting the first two equations ($a - d - (b - d) = 6 - 4$) tells us that A is 2 points per better than B. Similarly, A is 4 points better than C. This is true for any and all choices we make for d. So all of the infinity of ratings give the same ranking and point differentials. In particular, choosing $d = 0$ gives us $c = 2$, $b = 4$, and $a = 6$. Thus, the numbers in the far right column of the reduced matrix can be used as ratings!

Connected Schedules

The results from this example are typical. The reduced matrix has a row of 0's and the next-to-the-last column of the matrix is otherwise all -1's. The numbers in the last column can be used as Massey point ratings. The exception to this pattern is when the schedule is not **connected**.

In the above example, we have a basis for comparing teams A and C even though they have not played: A played B, which played C. Teams A and C are connected, even though they did not play. A schedule is connected if you can put together a chain of games connecting any two teams: A played B, which played C, which played .... It does not matter who won these games.

As long as the schedule is connected, the matrix will reduce as above and we get our ratings. Most leagues' schedules (even college football or basketball) become connected after a small number of games.

Applying the Massey rating system to 250 Division 1 college football teams in the 2014 regular season, we get the top five shown to the right. This list, unfortunately, fails the eye test for reasonableness. Undefeated Florida State is nowhere to be found (they were #19). How could the only undefeated team in the country not make the top ten? The answer is to think about how the system works. It uses points and schedules, period. If we want the ratings to take into account wins and losses, we need to incorporate wins and losses into the system. The value of a rating system depends on its underlying objectives!

1	Alabama	69.2
2	TCU	68.4
3	Oregon	65.2
4	Georgia	65.1
5	Ole Miss	64.7

TABLE 5.3: Top 5 Points

Massey Win Ratings

To have the Massey system use wins and losses, we can simply replace net points with net wins (wins minus losses). In effect, we declare each game to have a 1-0 outcome. Then proceed as before, reducing the matrix and reading off the win ratings from the far-right column. The matrix for our four-team league is shown to the right.

$$\begin{bmatrix} 5 & -2 & 0 & -3 & 5 \\ -2 & 4 & -2 & 0 & 0 \\ 0 & -2 & 4 & -2 & 0 \\ -3 & 0 & -2 & 5 & -5 \end{bmatrix}$$

The reduced matrix to the right gives us the win ratings, in this case $a = 15/11$, $b = 10/11$, $c = 5/11$, and $d = 0$. The ranking of A,B,C,D is the same as from the points ratings. However, it is not obvious what the values of the ratings represent. Properly scaled, the win ratings can be interpreted as points ratings. Multiplying by 22/5, the ratings become $a = 6$, $b = 4$, $c = 2$, and $d = 0$. More typically, the win ratings are distinct from the points ratings.

$$\begin{bmatrix} 1 & 0 & 0 & -1 & 15/11 \\ 0 & 1 & 0 & -1 & 10/11 \\ 0 & 0 & 1 & -1 & 5/11 \\ 0 & 0 & 0 & 0 & 0 \end{bmatrix}$$

The top five in win ratings from the 2014 college football regular season is quite different from the points ratings seen earlier. Notice, in particular, that unbeaten Florida State is now number one. While the points ratings penalized Florida State for not winning its games by large margins, the win ratings reward its undefeated season.

1	Florida St.	11.8
2	Alabama	11.8
3	Oregon	11.7
4	Ohio State	11.7
5	TCU	11.6

TABLE 5.4: Top 5 Wins

Offense and Defense

A nice feature of Massey's rating system is that the points ratings can be split into separate ratings for offense and defense.

Model: A team's points rating equals the sum of its offensive and defensive ratings.

In a game between teams A and B, team A's offensive rating minus team B's defensive rating gives the expected number of points for team A. Suppose that team A has a points rating of 64 and team B is rated at 54. On a neutral field, then, team A would be predicted to win by 10 points. Now, suppose that team A's rating breaks down as 40 on offense and 24 on defense (adding to 64), and team B's ratings are 38 on offense and 16 on defense. Team A should score $40 - 16 = 24$ points to team B's $38 - 24 = 14$ points.

Recall that the equations for the points (or wins) ratings do not have a unique solution; we have arbitrarily set the final team's rating to 0 in our examples. This lack of a unique solution carries over to the equations for offense/defense ratings, unless we add another equation to guarantee a unique solution (this, in fact, is what Massey did in his original work). To give the offensive and defensive ratings simple interpretations, the extra equation we add forces the sum of the defensive ratings to be 0. The top five offensive and defensive teams in the 2014 college football regular season are shown to the right. Because the defensive ratings sum to 0, the average defensive rating is 0. A team's offensive rating is therefore the number of points that team would score against an "average" defense. The defensive rating is the difference in the number of points that an opponent would score against that team compared to an "average" defense. The quotes are to bring attention to the fact that *average* depends on which teams are included in the ratings. In this case, all 250 Division 1 football programs were included. By Alabama's standards, an average Division 1 school is not very good.

1	TCU	56.1
2	Baylor	55.7
3	Oregon	55.0
4	Ohio State	54.4
5	Georgia	53.0

TABLE 5.5: Top 5 Offense

1	Ole Miss	25.9
2	LSU	24.6
3	Stanford	24.0
4	Alabama	22.9
5	Arkansas	21.0

TABLE 5.6: Top 5 Defense

Least Squares Equivalence

Kenneth Massey actually started with a different model for his system. Suppose that in the first game of the year team A beat team B by 3 points. We would like to have the ratings' prediction of $a-b$ match the actual outcome of 3. However, if in the second game between these teams A won by 2, we have a problem. We can't have $a - b$ equal to both 3 and 2.

The solution is to acknowledge that the predictions can't all be perfect, but our goal can be to minimize the errors. For the two games mentioned above, the errors would be $(a - b) - 3$ and $(a - b) - 2$. For several reasons, it is common to use the sum of the *squares* of the errors. For these two games, that would be $(a - b - 3)^2 + (a - b - 2)^2$. The values of a, b and so on that minimize the sum of the squares of the errors will be our ratings.

It turns out that the least squares solution for this problem is identical to the Massey point system described above. Thus, the Massey system makes sense on the global level (matching net points for the season) and the local level (giving the most accurate game-by-game predictions).

Wins versus Points

We have two different Massey systems; which is better? Recall that "better" depends on the objectives. Let's take as our main objective the most accurate prediction of future games (although the eye test of matching public opinion is also of concern). For convenience, we use "pr" for the point ratings and "wr" for the win ratings.

The table to the right shows the success rate of the two rating systems during the 2014 college football season. The favorite (the team predicted to win by the point spread) won 74.5% of the time. The Massey points system predicted winners 75.8% of the time, and beat the point spread 54.1% of the time. The results are good if not spectacular, especially for a system that utilizes such a small amount of information. Note the odd result that the point ratings did better picking winners, whereas the win ratings did better against the spread.

	Win %	v.Spread
pr	75.8	54.1
15 wr	73.2	55.1
spread	74.5	-

It seems reasonable to combine the two ratings. Here is one way to do so. First, put the ratings on a similar scale: 15 wr does a good job of predicting scores (55.1% against the spread in 2014). Then find the right proportion of each rating, such as 60% points and 40% wins. In other words, we want a combination $a\,\text{pr} + b\,(15\,\text{wr})$ with $a + b = 1$. Linear regression gives $a \approx .6$ as the right choice for the 2014 season. Then $\text{cr} = .6\,\text{pr} + 9\,\text{wr}$ did a good job of predicting college football games in 2014. The table shows its record, along with that of $\text{mr} = 4\,\text{pr} + 12\,\text{wr}$, the combination that I have used for years.

	Win %	v.Spread
cr	72.3	56.4
mr	74.5	56.4

The statistically inclined should note that the point and win ratings are highly correlated ($\rho = .98$), making multiple regression risky. However, pr and 15 wr−pr are only weakly correlated ($\rho = -.06$), so a regression of the form pr + c(15 wr−pr) is reasonable.

The Colley System

Wesley Colley's rating system was a mainstay of college football's BCS system. Its origins date back to the mathematician Pierre-Simon Laplace (1749-1827). When two competitors play for the first time, an unbiased estimate would give each a probability of $\frac{1}{2}$ of winning. If A beats B in round one, we might update the probabilities to something like $\frac{2}{3}$ for A. Laplace suggested that after w wins in g games, the formula $\dfrac{1+w}{2+g}$ gives a reasonable probability.

Note that when no games have been played the formula gives the desired probability of $\frac{1}{2}$. With successive wins, the formula approaches but never reaches 1, whereas with successive losses the formula approaches but never reaches 0. Also, it is common to treat a tie as half of a win and half of a loss.

Objective: A team's rating is based on Laplace's probability $\dfrac{1+w}{2+g}$ with w modified to include strength of schedule.

Colley's inspiration was to rework Laplace's probability to incorporate schedule. First, rewrite w as $(w-l)/2+(w+l)/2$. Now, if a team played an average schedule, then the average rating of its opponents would be $\frac{1}{2}$ and the sum of its opponents' ratings would be $g/2=(w+l)/2$.

Model: A team with w wins in g games has a rating of $\dfrac{1+w}{2+g}$ with w replaced by $(w-l)/2$ plus the sum of its opponents' ratings.

To see how this works, let's return to our four-team league. Team A has $w-l=5$, $g=5$, and opponent ratings that sum to $2b+3d$. Then

$$a=\frac{1+\frac{5}{2}+2b+3d}{7}.$$

Multiply across by 7 and move the b and d terms to the left to rewrite this equation as

$$7a-2b-3d=1+\frac{5}{2}=\frac{7}{2}$$

which is similar to the Massey win rating equation of $5a-2b-3d=5$. The team's rating is multiplied by $g+2$ instead of g, and the right-hand side is $1+(w-l)/2$ instead of $w-l$ (net wins).

$$\begin{bmatrix} 7 & -2 & 0 & -3 & 3.5 \\ -2 & 6 & -2 & 0 & 1 \\ 0 & -2 & 6 & -2 & 1 \\ -3 & 0 & -2 & 7 & -1.5 \end{bmatrix}$$

Although they start with very different models, the Massey and Colley systems end up with similar equations. The small differences in equations make a large difference in their solution. The Colley system has a unique solution. In this case, we get (rounded to 2 decimal places) $a=.76$, $b=.57$, $c=.43$, and $d=.24$.

Notice that the average rating in the above example is .5. You will show in the exercises that this is always the case. An improvement in one team's rating is accompanied by equal declines in others.

As formulated, the Colley system uses only schedules and won/lost records. The similarity in matrix equations of the Massey win and Colley systems gives us a way to adjust the Colley system to use points. Replace the right-hand side of the Colley equations with net points, and solve the system. For our example, we get rounded values of $a = 2.32$, $b = .58$, $c = -.58$, and $d = -2.32$. The teams are separated by fewer points in this system than in the Massey point system, where the gaps between teams were a consistent 2 points.

A Flaky Scaling Problem

There are several properties that it just seems reasonable for a ranking system to possess. Suppose a league plays a complete season and you apply a ranking system. Then, in the ultimate deja vu, the season gets replayed with exactly the same result. Instead of team A beating team B by 12 points, we now have team A beating team B by 12 points *twice*. It seems silly to run the ranking system on this double season, right? Except that the Colley system changes!

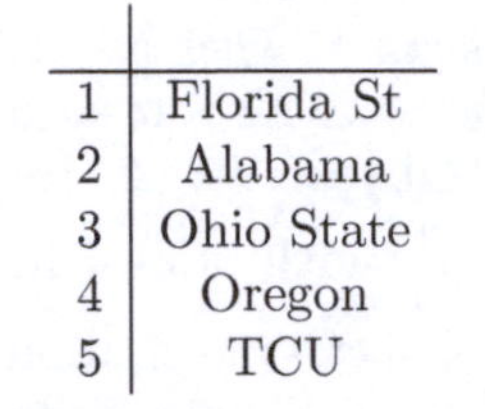

1	Florida St
2	Alabama
3	Ohio State
4	Oregon
5	TCU

TABLE 5.7: Colley Top 5

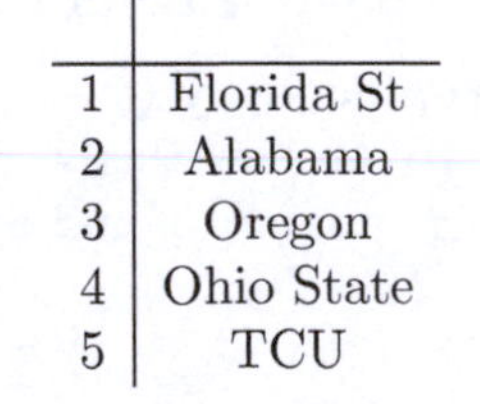

1	Florida St
2	Alabama
3	Oregon
4	Ohio State
5	TCU

TABLE 5.8: Colley Top 5

To the right are the Colley ratings for the 2014 college football regular season in Table 5.7, and the Colley ratings for the 2014 double season in Table 5.8. Ohio State and Oregon switch positions! This is not an isolated phenomenon: only 7 of the top 20 teams maintained their original ranks, with teams jumping up and down by as many as 3 positions.

How weird is this? Recall that the Colley system is based on the Laplace formula $\frac{1+w}{2+g}$. If a team wins 3 out of 5 games, we update their Laplace probability to $\frac{4}{7}$. If the team wins 3 of their next 5 games, we update the probability to $\frac{7}{12}$. This does not seem odd at all. If the team consistently wins 3 out of 5 games, we want our rating to converge to the probability $\frac{3}{5}$ from its starting point of $\frac{1}{2}$. You can check that $\frac{7}{12}$ is closer to the destination $\frac{3}{5}$ than is $\frac{4}{7}$. Perhaps it is not illogical for the ratings to change, even if it is disconcerting that the rankings would change.

The Elo System

Arpad Elo (1903-1992) developed a system to rate chess players. His system, slightly modified, is still used to compute the official chess ratings. More recently, Jeff Sagarin's adaptation of the Elo system to rate college football teams was a part of the BCS system.

Objective: A team's rating is updated after every game based on its performance in that game, with the change in rating reflecting the result and the quality of the opponent.

Think about a league in which every team plays on given days (for example, once a week). After each round of games, the Massey and Colley systems essentially start from scratch, compiling the schedules and records of each team for the entire season. By contrast, the Elo system takes the previous ratings and makes adjustments based on the most recent results. The rating of a team that just won increases, while the rating of a team that lost decreases.

Model: A team's rating r_{new} equals its previous rating r_{old} plus an adjustment $k(s-m)$, where k is a constant, s is the team's performance in its last game, and m is its predicted performance in that game.

We will explore each part of this model in turn. First, you need a set of ratings to adjust once games are played. You could be unbiased and give each team an equal initial rating, or you could use the best prediction available to seed the ratings (this could be records from the previous year, or some other form of preseason rating).

The constant k controls how much the ratings will change based on the outcome of one game. For chess, the values of k range from 25 for new players down to 10 for experienced experts. For sports like soccer or tennis, different values of k can be used to indicate the importance of a match (e.g., 60 for a World Cup match down to 20 for a friendly). A value of $k = 32$ has worked well for the NFL.

The game performance s can indicate winners or point spreads. In chess, $s = 1$ for the winner, $s = 0$ for the loser, and both players get $s = .5$ for a draw. For a sport with points, Langville and Meyer suggest $s = \dfrac{pf+1}{pf+pa+2}$ for a team that scored pf points and allowed pa points. Note the use of the Laplace formula, so that $0 < s < 1$ and, more importantly for the properties of the system, the sum of the s values for the two opponents equals 1.

The predicted performance m is intended to quantify the gap in abilities of the two opponents as essentially a probability that each player will win. The value suggested by Elo was

$$m = \frac{1}{1 + 10^{-(r_a - r_b)/400}}$$

for team A entering the game with rating r_a against team B rated at r_b. This may look intimidating, but Figure 5.1 shows that m increases smoothly from 0 to 1 as the difference in rating goes from $-\infty$ to ∞.

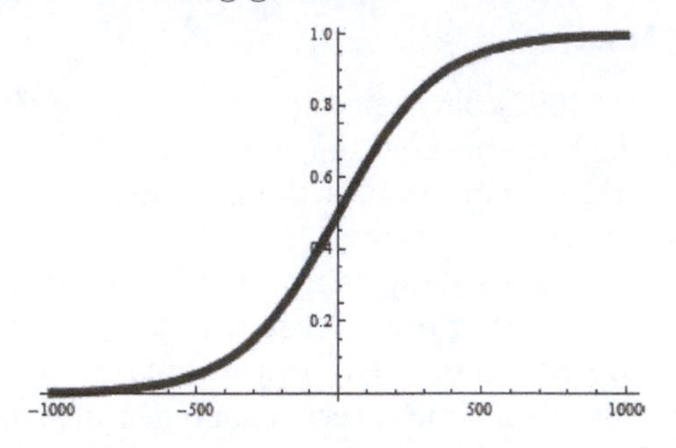

FIGURE 5.1: Elo Performance Graph

An example should help. Person A enters a chess match with rating 2600, facing person B whose rating is 2200. Based on the ratings, we would have no trouble picking A to win. For player A, we compute $m = 1/(1+10^{-400/400}) = 1/(1+0.1) = 0.91$. We would predict A to win 10 out of 11 matches, or 91%. For player B, we compute $m = 1/(1+10^{--400/400}) = 1/(1+10) = 0.09$. We predict B to win 9% of the time. Note that the probabilities sum to 1.

Now, suppose B wins in an upset. We want B's rating to increase and A's to decrease, but by how much? If A and B were young players, we might want B's new rating to be higher than A's. If they had played often (say, 11 times, with A winning 10), we would not want B to jump above A. Taking the value of k to be 10, we compute the new ratings: $b = 2200+10(1-.09) = 2209.1$ and $a = 2600 + 10(0 - .91) = 2590.9$. Notice that B's rating increased by 9.1 and A's decreased by the same amount. This symmetry is built into the system.

This example should make the significance of the choice of k clearer. Since $s - m$ will always be between 0 and 1, k is the maximum number of points that a rating can change. For a given sport, its value should be adjusted so that the ratings are stable (don't change too much) but responsive (reflect changes in performance).

Notice also that the 400 in the exponent plays an important role in controlling the rate at which the probabilities approach 0 and 1. If 400 is changed to 1600 in the above calculation, the probability that A wins drops to 0.64. A ratings difference of 1600 would be required to reach a win probability of 0.91. This is explored further in exercise 5.17.

Strength of Schedule

Each of the systems presented in this chapter incorporates strength of schedule. This is a critical feature of any rating system. Most people have a good intuitive understanding of what the concept of strength of schedule represents. However, a precise determination of how to calculate strength of schedule turns out to be surprisingly difficult.

Think about teams A and B, each of which plays 10 games. Team A's opponents are all good, but not great, teams. Team B plays the three best teams, three mediocre teams, and four bad teams. Which team has a tougher schedule? Depending on which system you use to rate teams, team A's opponents may have a higher average than team B's opponents. However, there are other ways to measure strength of schedule.

If both A and B are mediocre to good teams, then A's schedule is loaded with maybe win/maybe lose games, while B has 4 sure wins and 3 more easy games. A reasonable prediction might be 7 wins for B against 5 wins for A. So A's schedule looks tougher.

However, what if A and B are both great teams? While A's games are all losable, A would be heavily favored in every game; 10-0 and 9-1 are reasonable predictions. B, however, has three very tough games and 7-3 and 8-2 results are very possible. Now B's schedule looks tougher.

Both arguments are valid. If you are trying to pick the top four teams for a playoff, however, the second argument should feel more convincing. But here we run into more complications. For the top teams, does it make more sense to compute an expected record or to compute the probability of going undefeated?

The bottom line here is that strength of schedule is one of those terms that sounds simple enough that we don't necessarily question the assumptions that go into its calculation. When you see a strength of schedule rating, you should try to find out how it is computed.

Computing Probabilities

One way to compute strength of schedule is to calculate the probability that a team of a certain strength could go undefeated. This, of course, requires that we compute the probabilities that teams of certain strengths will win particular games. If a team with rating 100 faces a team with rating 94, what is the probability that the better team wins?

Phrasing the question in this way may prompt you to ask an important question. How much luck is there in the game? One team is considered to be

6 points better, but is that significant? Six goals in soccer or hockey are hard to score, but six points in football can easily result from a weird bounce of the ball.

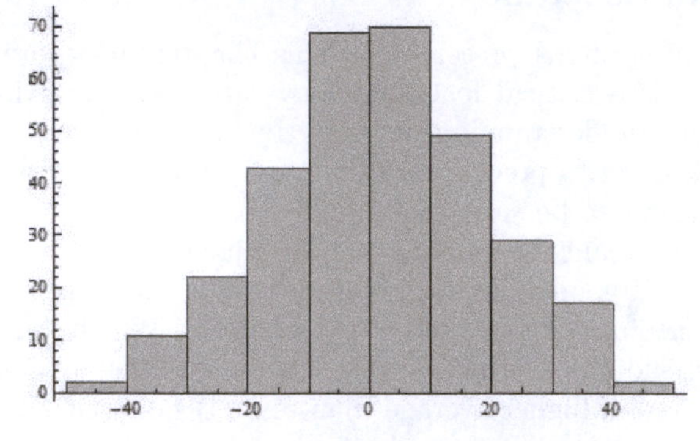

FIGURE 5.2: Predicted Minus Actual Scores, 2014

We will explore this question more fully in Chapter 8, but for now we give one answer. For the 2014 college football regular season, the Massey ratings (actually the combination ratings named "mr" above) were used to predict all Division 1-A games. The histogram in Figure 6.2 shows the frequencies of values of predicted score minus actual score. The plot looks like a nice bell curve. The mean is essentially 0, and the standard deviation is about 16 points. That is, start with the Massey prediction and nearly one-third of the games ended more than 16 points different than the prediction (including two that were more than 40 points different)! That is a lot of plus/minus. In that context, a 6-point difference does not seem very large. If the game score differences are really normally distributed with mean 0 and standard deviation 16, the probability that a 6-point favorite wins is approximately 64.6%. (The probability that a 16-point favorite wins would compute to be 84%; does that surprise you?)

Weighty Issues

One objection to the Massey and Colley systems as described here is that all games are weighted equally. Games that are played early in the season count as much as games that are played at the end of the season. Thus, a team that improves during the season will be performing at a higher level than its rating indicates. In some sports (soccer, golf, tennis, and others), not all contests have equal importance. Matches in a major tournament should count more than glorified scrimmages that no one cares about.

Fortunately, both the Massey and Colley systems are easily tweaked to count the most important games extra. In particular, there is no requirement that the number of games be an integer. You could count an especially important game (late in the season, or in a major tournament) as 1.5 games, or an unimportant game as 0.7 games. Winning a 0.7 game would give you 0.7 wins, or winning a 0.7 game by 10 points would give you 7 net points.

What is the best way to weight games? You should not be surprised to read that "best" depends on your objectives. In a 16-game season, do you want all games to count equally or not? If not, should game 16 count more than game 15, or should games 13-16 count the same (and more than games 9-12)? Here are three types of functions you might use for weighting games in time.

Linear: The weight function is a line. If the last of n games counts as b games and the first as a games, then the i-th game counts $a + (i-1)(b-a)/(n-1)$ games. Every game is given a different weight, with the difference in weights between the 2nd and 3rd games equal to the difference in weights between the 15th and 16th games.

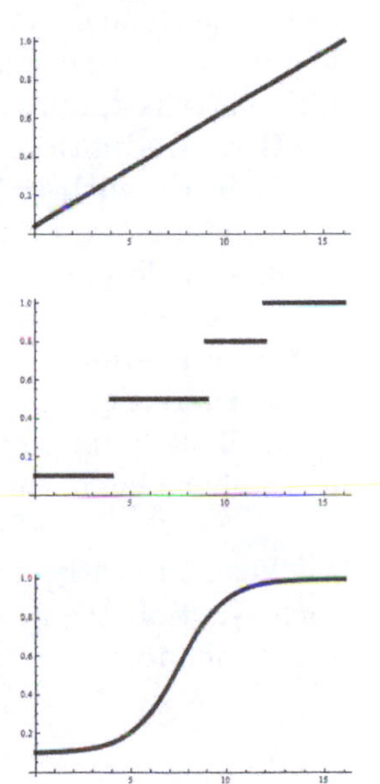

Step: The weight function is a sequence of stair steps. The last n_1 games count as b_1 games, the n_2 games before that as b_2, and so on. In the figure shown, the stairsteps are not equally-sized. You can weight groups of games as you wish. Here, the last 4 games all carry equal weight.

Logistic: The weight function increases in an S-shape. To increase from a games to b games, the i-th game counts $a + \dfrac{b-a}{1+ce^{d(i-w)}}$ for constants c, d, and w that can be adjusted. While all games have different weights, the last few games are weighted nearly equally, as are the first few games. There is a large change in the weights in the middle games.

The proper choice of weights depends on the details of the sport and the objectives of the ratings system.

Calculus Box: A Recipe for Reduction of Matrices

Technically, this is a *linear algebra* box to demonstrate a method for finding the reduced form of a matrix. There are numerous calculators and software packages that will do this for you. As you will see, the method is not complicated, but requires extensive arithmetic; in other words, it is an ideal computer application.

$$\begin{bmatrix} 5 & -2 & 0 & -3 & 22 \\ -2 & 4 & -2 & 0 & 0 \\ 0 & -2 & 4 & -2 & 0 \\ -3 & 0 & -2 & 5 & -22 \end{bmatrix}$$

Let's start with the matrix for the Massey points system for the league in Tables 5.1 and 5.2. The *elementary row operations* we will use are based on equation solving techniques you have used before. For example, if you needed to solve the system of two equations and two unknowns with $x - y = 2$ and $3x + 2y = 11$, the technique of *elimination* can be used. Multiply the first equation by 2 to get $2x - 2y = 4$ and then add it to the second equation $3x + 2y = 11$. We get $2x - 2y + 3x + 2y = 4 + 11$ or $5x = 15$, from which we conclude $x = 3$. Equivalently, we can eliminate a non-zero element from a row of a matrix by adding a constant times one row to another row. The algorithm that follows using the standard matrix notation that the (i,j) entry is the number in the i-th row and j-th column. For example, in our matrix the (1,5) entry is 22, the (2,2) entry is 4, the (3,2) entry is −2, and so on.

Row Reduction Algorithm:

1. Start with i=1.

2. Make the (i,i) entry nonzero, swapping rows if necessary.

3. Divide each number in the i-th row by the (i,i) entry.

4. For each j≠i, make the (j,i) entry 0, replacing row j with the old row j minus the old (j,i) entry times row i.

5. Increase i by 1 and repeat steps 2-4 until finished.

Looking back at our matrix, there is a 5 in the (1,1) entry, so step 3 tells us to divide by 5. The first row is now

$$\begin{bmatrix} 1 & -2/5 & 0 & -3/5 & 22/5 \end{bmatrix}$$

The (2,1) entry and (4,1) entry are not yet zero, so we apply step 4 to each. Multiply the new row 1 by the old (2,1) entry (−2) and subtract it from the second row. That is,

	−2	4	−2	0	0
−	−2	4/5	0	6/5	−44/5
=	0	16/5	−2	−6/5	44/5

Our goal of a zero in the first slot is accomplished. We need to do the same for the fourth row. Take row 4 minus −3 times row 1 (this is the same as row 4 plus 3 times row 1).

	−3	0	−2	5	−22
+	3	−6/5	0	−9/5	66/5
=	0	−6/5	−2	16/5	−44/5

$$\begin{bmatrix} 1 & -2/5 & 0 & -3/5 & 22/5 \\ 0 & 16/5 & -2 & -6/5 & 44/5 \\ 0 & -2 & 4 & -2 & 0 \\ 0 & -6/5 & -2 & 16/5 & -44/5 \end{bmatrix}$$

This completes step 4. The updated matrix is shown to the right. Moving on in the algorithm to i = 2, we implement step 3 by dividing the second row by 16/5. This is the same as multiplying by 5/16.

The matrix with its new second row is shown to the right. We will need to execute step 4 three times, to create zeros in the first, third, and fourth rows. Letting the computer do this is looking like a good idea. For j=1, we take the first row plus 2/5 times the second row.

$$\begin{bmatrix} 1 & -2/5 & 0 & -3/5 & 22/5 \\ 0 & 1 & -5/8 & -3/8 & 11/4 \\ 0 & -2 & 4 & -2 & 0 \\ 0 & -6/5 & -2 & 16/5 & -44/5 \end{bmatrix}$$

	1	−2/5	0	−3/5	22/5
+	0	2/5	−1/4	−3/20	11/10
=	1	0	−1/4	−3/4	11/2

Notice that we retain the one in the first slot. Since we are working on the second column (i=2) the desired 0 is in the second column. The next step is row 3 plus 2 times row 2.

	0	−2	4	−2	0
+	0	2	−5/4	−3/4	11/2
=	0	0	11/4	−11/4	11/2

Finally, we take row 4 plus 6/5 times row 2.

	0	−6/5	−2	16/5	−44/5
+	0	6/5	−3/4	−9/20	33/10
=	0	0	−11/4	11/4	−11/2

This completes step 4. The updated matrix is shown to the right. We now move on to i = 3. To create a 1 in the (3,3) slot, we implement step 3 by dividing the third row by 11/4. This is the same as multiplying by 4/11.

$$\begin{bmatrix} 1 & 0 & -1/4 & -3/4 & 11/2 \\ 0 & 1 & -5/8 & -3/8 & 11/4 \\ 0 & 0 & 11/4 & -11/4 & 11/2 \\ 0 & 0 & -11/4 & 11/4 & -11/2 \end{bmatrix}$$

The matrix with its new third row is shown to the right. We again have three zeros to create, two above and one below the (3,3) entry. The work is shown below, starting with row 1 plus 1/4 times row 3.

$$\begin{bmatrix} 1 & 0 & -1/4 & -3/4 & 11/2 \\ 0 & 1 & -5/8 & -3/8 & 11/4 \\ 0 & 0 & 1 & -1 & 2 \\ 0 & 0 & -11/4 & 11/4 & -11/2 \end{bmatrix}$$

	1	0	−1/4	−3/4	11/2
+	0	0	1/4	−1/4	1/2
=	1	0	0	−1	6

We next want row 2 plus 5/8 times row 3.

	0	1	−5/8	−3/8	11/4
+	0	0	5/8	−5/8	5/4
=	0	1	0	−1	4

Finally, we take row 4 plus 11/4 times row 3.

	0	0	−11/4	11/4	−11/2
+	0	0	11/4	−11/4	11/2
=	0	0	0	0	0

Putting this together gives us the desired reduced matrix.

$$\begin{bmatrix} 1 & 0 & 0 & -1 & 6 \\ 0 & 1 & 0 & -1 & 4 \\ 0 & 0 & 1 & -1 & 2 \\ 0 & 0 & 0 & 0 & 0 \end{bmatrix}$$

Exercises

In these exercises, (T) refers to thinking problems, conceptual problems requiring no calculations. (C) refers to problems requiring calculus, linear algebra or significant computer calculations. (P) refers to projects; these are ideas for further investigation (hints and resources are at the book's web site).

5.1 Suppose that

Team	W	L	PF	PA
A	5	0	74	50
B	2	3	58	53
C	3	2	52	47
D	0	5	40	74

and

	A	B	C	D
A	-	2	1	2
B	2	-	2	1
C	1	2	-	2
D	2	1	2	-

give a league's results and schedule. Set up the matrices needed to find the (a) Massey win ratings; (b) Massey points ratings; (c) Colley ratings.

5.2 The reduced matrices for exercise 5.1 round to (a) $\begin{bmatrix} 1 & 0 & 0 & -1 & 1.42 \\ 0 & 1 & 0 & -1 & 0.67 \\ 0 & 0 & 1 & -1 & 0.75 \\ 0 & 0 & 0 & 0 & 0 \end{bmatrix}$; (b) $\begin{bmatrix} 1 & 0 & 0 & -1 & 8.46 \\ 0 & 1 & 0 & -1 & 6.5 \\ 0 & 0 & 1 & -1 & 5.29 \\ 0 & 0 & 0 & 0 & 0 \end{bmatrix}$; (c) $\begin{bmatrix} 1 & 0 & 0 & 0 & .775 \\ 0 & 1 & 0 & 0 & .475 \\ 0 & 0 & 1 & 0 & .525 \\ 0 & 0 & 0 & 1 & .225 \end{bmatrix}$. Find the ratings and rankings for the three systems, and comment on any interesting features. Determine a scaling factor c such that c times the Massey win ratings are of the same magnitude as the Massey point ratings. Give one reason why you can know in advance that, unlike the example in the text, there is no c that will produce an exact match.

5.3 Suppose that one more game is played in the league of Tables 5.1 and 5.2, with C beating A by a score of 12-6. Set up the matrices needed to find the (a) Massey win ratings; (b) Massey points ratings; (c) Colley ratings.

5.4 The reduced matrices for exercise 5.3 round to (a) $\begin{bmatrix} 1 & 0 & 0 & -1 & 1.13 \\ 0 & 1 & 0 & -1 & 0.97 \\ 0 & 0 & 1 & -1 & 0.81 \\ 0 & 0 & 0 & 0 & 0 \end{bmatrix}$; (b) $\begin{bmatrix} 1 & 0 & 0 & -1 & 4.75 \\ 0 & 1 & 0 & -1 & 4.31 \\ 0 & 0 & 1 & -1 & 3.88 \\ 0 & 0 & 0 & 0 & 0 \end{bmatrix}$; (c) $\begin{bmatrix} 1 & 0 & 0 & 0 & .67 \\ 0 & 1 & 0 & 0 & .57 \\ 0 & 0 & 1 & 0 & .54 \\ 0 & 0 & 0 & 1 & .23 \end{bmatrix}$. Find the ratings and rankings for the three systems, and comment on the difference that one game can make early in a season.

5.5 Set up the (nine) equations for the Massey offensive and defensive ratings for the league in Tables 5.1 and 5.2. In this case, the ratings are the "least squares solution" of the equations. (See exercise 5.34 below.)

5.6 The solutions for exercise 5.5 (in order for teams A, B, C, and D) are offensive ratings of 14.46, 10.93, 10.07, and 7.82 and defensive ratings of −.64, .89, −.25, and 0. Verify that the defensive ratings sum to 0 and compute the point ratings as the sum of the offensive and defensive ratings. The point ratings are different from those reported in the text. Explain the differences.

5.7 For the teams of exercises 5.5 and 5.6, predict the scores of games between (a) A and B; (b) A and C; (c) A and D; (d) B and C; (e) B and D; (f) C and D. Note that game scores should be integers.

5.8 Suppose that

Team	W	L
A	2	0
B	0	2
C	1	1
D	1	1

and

	A	B	C	D
A	-	2	0	0
B	2	-	0	0
C	0	0	-	2
D	0	0	2	-

give a league's results and schedule. Explain why this league's schedule is not connected. At this point, is there any basis for comparing, for example, teams A and C? Set up the matrices needed to find the (a) Massey win ratings; (b) Colley ratings.

5.9 The reduced matrices for exercise 5.8 are (a) $\begin{bmatrix} 1 & -1 & 0 & 0 & 1 \\ 0 & 0 & 1 & -1 & 0 \\ 0 & 0 & 0 & 0 & 0 \\ 0 & 0 & 0 & 0 & 0 \end{bmatrix}$; (b) $\begin{bmatrix} 1 & 0 & 0 & 0 & 3/4 \\ 0 & 1 & 0 & 0 & 1/4 \\ 0 & 0 & 1 & 0 & 1/2 \\ 0 & 0 & 0 & 1 & 1/2 \end{bmatrix}$. For part (a), explain why this is not the standard form for a reduced Massey matrix. Translate the matrix back into equation form to interpret the results. Explain why these ratings make sense. For part (b), read off the Colley ratings. Explain why these ratings make sense within the context of the model for the Colley system.

5.10 In the league of exercise 5.8, if teams A and C play then the schedule becomes connected. Suppose that C beats A. Set up the matrices needed to find the (a) Massey win ratings; (b) Colley ratings. (c) What do you expect to happen to the ratings? Explain.

5.11 The reduced matrices for exercise 5.10 are (a) $\begin{bmatrix} 1 & 0 & 0 & -1 & -1 \\ 0 & 1 & 0 & -1 & -2 \\ 0 & 0 & 1 & -1 & 0 \\ 0 & 0 & 0 & 0 & 0 \end{bmatrix}$;

(b) $\begin{bmatrix} 1 & 0 & 0 & 0 & .53 \\ 0 & 1 & 0 & 0 & .27 \\ 0 & 0 & 1 & 0 & .63 \\ 0 & 0 & 0 & 1 & .57 \end{bmatrix}$. Find the ratings and rankings for the two systems and discuss the results. In particular, does it make more sense to you to have teams C and D tied or to have C above D?

5.12 Look up points scored and allowed per game for the 2014 college football regular season. How do the top five in Tables 5.5 and 5.6 compare? Explain why there are differences between the ratings and the actual point totals.

5.13 The Massey point ratings can be tweaked to determine home court advantage. Suppose a league has the following results after a home-road round-robin in which every team plays each of the other teams twice, once at each team's home court. Team A scored 34 points more than its opponents in the three games it played at home, but scored 10 points less than its opponents in the three games it played on the road.

Team	HPD	RPD
A	34	-10
B	12	3
C	-3	-20
D	4	-20

Consider each team as being different when playing at home and on the road. Thus, there is team AH (team A playing at home) and a separate team AR (team A playing on the road). Set up the matrix for the Massey point ratings for the *eight* teams in this league. The ratings for the home teams end up being 14.2, 4.3, 1.5, and 4.2 and the road ratings are 0, 7.6, 0.9, and 0. Discuss any interesting features. In particular, why is BR rated above BH even though its point differential of 3 is worse than BH's point differential of 12? Why are AR and DR rated the same? To compute each team's home court advantage, subtract its road rating from its home rating. Discuss any interesting features.

5.14 Show that the Colley ratings will always average 1/2. (Hints: add the columns of the Colley matrix and convert this into an equation. Simplify the equation to conclude that the sum of the ratings equals $n/2$, where n is the number of teams in the league.)

5.15 Show that the average Elo rating is constant as long as the sum of the results $s_1 + s_2$ and the sum of the expected results $m_1 + m_2$ both equal 1. (Hint: compute the two changes in ratings resulting from a match.)

5.16 Experienced chess players with ratings 2800 and 2600 play each other. Compute the new ratings if (a) the better player wins; (b) the lesser player wins.

5.17 For chess ratings, show that every 400 points difference indicates that a player is 10 times better than the opponent. To be precise, if A is ranked $400n$ points higher than B for some positive integer n, show that the probability of A winning is 10^n times the probability of B winning.

5.18 If experienced chess players A and B both start with ratings of 2600, A beats B, and then B beats A, are A and B rated the same? Should they be?

5.19 The USGA golf handicap system uses the average of the best 10 of the last 20 rounds. For a golfer whose last 20 rounds, in order, produce adjusted scores of 6, 11, 8, 4, 14, 9, 6, 8, 7, 8, 8, 11, 13, 10, 9, 8, 12, 9, 7, and 9 (small is better) compute the average. What should happen if the golfer's next score is 7? What does happen? What happens if the score after that is 16? Comment on this system.

5.20 Elo ratings for the NFL use $k = 32$ and an exponent of $-(r_a - r_b)/1000$. If A is rated 110 and B is rated 100, update the ratings if (a) A wins; (b) B wins.

5.21 Repeat exercise 5.20 replacing the 1000 in the exponent with (a) 1500; (b) 500. Which value in the exponent is most reasonable? Explain.

5.22 (T) Two teams finish the season with the same record. They played each other once. Do you think that the winner should definitely be ranked above the loser? Give arguments both for and against.

5.23 (T) Most sports leagues are divided into conferences and/or divisions. Suppose that a team that does not win its conference/division makes the playoffs and wins the league championship. Does that indicate that the regular season is meaningless or that the playoffs are highly random? Discuss.

5.24 (T) If a league wanted to choose the "best" teams (the ones playing the best at the end of the season) for the playoffs, how could it do so? Describe a modification of one of the ratings systems in this chapter that could be used.

5.25 (T) Voters in college sports polls are asked to rank teams from 1-25. Discuss advantages and disadvantages of allowing the voters to rate the top 25 compared to the top 1 or top 10.

5.26 (T) Label each of the following as a rating or ranking problem. (a) setting point spreads for upcoming games; (b) determining a conference champion; (c) determining seeds for a playoff; (d) the lists in the opening graphic for the chapter.

5.27 (T) Compare the models for the Massey, Colley, and Elo systems and state which one seems to be the most reasonable. Explain.

5.28 (T) Name several factors that could be important for rating teams but that are *not* used by the Massey, Colley, or Elo systems.

5.29 (T) The four-team 2014 college football playoffs included Alabama, Oregon, Florida State, and Ohio State. Based on Tables 5.3 and 5.4, was the selection committee more influenced by wins or points? Comment on whether the committee was more focused on the "right" teams or the "best" teams.

5.30 (T) If team B is a heavy underdog but upsets team A, should B's rating go up a lot more than A's rating goes down? Does the Elo system do this? Answer in terms of both numerical change and percentage change.

5.31 (T) Explain how strength of schedule is used in each of the Massey, Colley, and Elo systems.

5.32 (T) If a league's schedule and results are doubled (the scenario in "A Flaky Scaling Problem"), do the Massey ratings change? Explain. Do the Elo ratings change? Explain.

5.33 (T) Early in the season team A is outstanding, but several injuries in midseason cause A's quality to drop to mediocre by the end of the season. Should A's rating indicate a mediocre team or one that is halfway between outstanding and mediocre? Should teams that played A early in the season get credit for playing an outstanding team, or for playing one with A's rating at the end of the season? Discuss.

5.34 (C) Here is some of the linear algebra behind exercise 5.5. The left-hand sides of your nine equations can be formed into a 9x8 matrix M, with a 9x1 vector b formed by the right-hand sides. Ideally, we would just solve Mx = b for x, but there is no exact solution. We find the best match by multiplying the equation by M^T, the transpose of M. The solution of $M^T Mx = M^T b$ is our rating vector reported in exercise 5.5. Compute $M^T M$ and $M^T b$ and solve the system.

5.35 Ⓒ Explain in linear algebra terms why every column summing to 0 guarantees that the set of row vectors is linearly dependent. Explain in terms of schedules and results why the set of row vectors is linearly dependent. (Hint: if you know the schedules and results of all but one team, can you determine the schedule and results of the remaining team?)

5.36 Ⓒ Reduce the matrices of exercise 5.1 (a)-(c) to obtain the matrices shown in exercise 5.2.

5.37 Ⓒ Assuming a normal distribution of scores with mean 0 and standard deviation 16 and assuming all games are independent, find the probability that a team of rating 90 goes undefeated against each schedule. (a) 5 games against teams of rating 80; (b) games against teams of rating 92, 90, 88, 60, and 60.

5.38 Ⓒ For the schedules of exercise 5.37, (a) which one has the higher mean rating? (b) Which one is harder to go undefeated against? (c) Which one is harder to win at least 3 games against?

5.39 Ⓒ Pick your own sport and year and rate teams using the Massey point, Massey win, and Colley systems. An interesting version of this assignment is to take the schedules and records at the halfway mark of the season and use the first-half ratings to predict the eventual standings.

5.40 Ⓟ Develop your own rating system by modifying one of the systems in this chapter. Be explicit about the reasons for your modifications: what are your objectives and why are your changes likely to address them?

5.41 Ⓟ Choose a weighting system and apply it to the rating system and league of your choice. Explain your choice of weighting system.

5.42 Ⓟ Research the Keener method or some rating system not covered in this chapter and apply it to the examples in this chapter or to your favorite sport. Discuss the apparent advantages and disadvantages of this method.

Further Reading

Amy Langville and Carl Meyer's *Who's #1?* is an excellent resource for the ratings systems covered here plus several more.

Tim Chartier and Drew Pasteur contributed excellent chapters to *Mathematics and Sports* edited by Joe Gallian.

An impressive amount of information can be found at Kenneth Massey's www.masseyratings.com. There are also good web sites for the Colley, Elo, and Sagarin systems.

The UMAP Module referenced for the Massey system is number 725 "A Mathematical Rating System" by Roland Minton. Thomas Jech's 1983 *Americam Mathematical Monthly* article is what got me started. My "blogging" about my rating system can be found at my *By the Numbers* web site.

Chapter 6

Voting Systems

Introduction

Sports fans love to argue. We get especially engaged (and sometimes enraged) when the topic is ranking the best players and teams of all time. We take our Most Valuable Player (MVP) and Hall of Fame (HOF) voting seriously.

Both MVP and HOF voting will be featured in this chapter, but the primary focus is on the mathematics of voting. Whether Pete Rose or Barry Bonds deserves to be in the Baseball HOF in Cooperstown is a question for a different book, perhaps an ethics text. Whether Roger Maris belongs in Cooperstown is a question for a different chapter, where we look at baseball analytics. Here, we will analyze the various voting methods that are used to determine MVPs and HOFs.

A quick scan of sports stories at almost any time shows the importance of voting in sports. The Heisman Trophy in college football, the Baseball HOF, the MVPs in baseball and pro football, the NCAA basketball polls, and countless cell phone surveys all involve voting of one sort or another.

We will analyze many of the voting methods used in these cases, trying to determine what biases or injustices might be promoted by the voting rules. Among the questions we will answer are the following. Which types of voting systems are used? What are the flaws in these systems? Is there a best voting system? What do Google and Amazon have to do with voting? Why was Kenneth Arrow awarded a Nobel Prize for his contributions to voting theory? Is it better to be seeded ninth or tenth in the NCAA Basketball Tournament?

How They Vote

We start by reviewing some of the voting methods used in sports.

The Heisman Trophy is the most prestigious award in college football. The award is decided by a group of approximately one thousand voters, with each voter listing his or her top three players. Players receive 3 points for each first-place vote, 2 points for each second-place vote, and 1 point for each third-place vote. The player with the most points wins.

One of the closest Heisman votes occurred in 2008. Oklahoma's Sam Bradford won with 300 first-place votes, 315 second-place votes, and 196 third-place votes for $300(3) + 315(2) + 196(1) = 1726$ points. Colt McCoy of Texas was second with 266 first-place votes, 288 second-place votes, and 230 third-place votes for 1604 points, and Tim Tebow was third with 309 first-place votes, 207 second-place votes, and 234 third-place votes for 1575 points. Is there anything that bothers you about this? For some, the fact that Tebow had more first-place votes than Bradford means that Tebow should have won. A counterargument is that 811 voters had Bradford in the top three compared to 750 voters for Tebow.

The MVP voting in professional football will not have this issue. Voter Barry Wilner said, "We don't want an MVP who doesn't get the most choices, so we use a better system that guarantees the player with a majority of first-place votes wins." The NFL MVP is decided by 50 voters who each vote for one player. The player with the most votes wins. In most years, the winner does indeed have a majority of votes, but in 2005 Shaun Alexander won with 19 of the 50 votes (38%).

We need two definitions. A **majority** of votes is more than 50% of the votes. For the 50 votes in the NFL's system, a majority is 26 or more votes. (25 votes, exactly 50%, does not constitute a majority; only one person can have a majority of the votes.) Shaun Alexander did *not* have a majority of votes in 2005. Alexander, however, did have a plurality of votes: a **plurality** of votes is the most first-place votes received by any candidate. Most elections in the United States use the plurality system. When there are more than two strong candidates, the plurality winner could have a small number of votes. As we have seen, people do not always accurately distinguish between majority and plurality.

Baseball and basketball do not share the NFL's objection to the point system (or its confusion of majority and plurality). The major league baseball (MLB) MVP system asks each voter to list ten players in order, with players receiving 14 points for each first-place vote, 9 points for each second-place vote, 8 points for each third place vote, and so on down to 1 point for each tenth-place vote. Voters for the Cy Young Award for best pitcher list 5 players, who receive 7, 4, 3, 2, or 1 point for votes. Point systems like this are susceptible to the 2008 Heisman situation of the plurality winner not being the overall winner. In 1966, Roberto Clemente edged Sandy Koufax for National League MVP in spite of receiving only 8 first-place votes to Koufax's 9.

The NBA uses a point system for its MVP with the top five for each voter receiving 10, 7, 5, 3, or 1 point for votes. In 1990, Magic Johnson was elected MVP despite receiving only 27 first-place votes, compared to Charles Barkley's 38 first-place votes. Magic received 38 second-place votes to Barkley's 15 to more than make up the point difference.

The Baseball Hall of Fame uses a different type of voting system. From a list of eligible players, each of the approximately 500 voters may vote for up to ten players whom the voter deems worthy of inclusion. Any number of votes from zero to ten is allowed. To be elected, a player must be listed on at least 75% of all submitted ballots.

For the selection of the host city of an Olympic Games, cities are eliminated until one city receives a majority of first-place votes. In voting for the 1994 Winter Olympics, 84 voters each voted for one city. In the first round, Lillehammer received 25 votes, Anchorage 23, Ostersund 19, and Sofia 17. Since the plurality winner, Lillehammer, did not receive a majority of votes (25 out of 84 is less than 30%), a second round of voting was needed. The last-place city from the previous round, in this case Sofia, was dropped and another vote was taken. This time Ostersund received 33 votes (39%), Lillehammer 30, and Anchorage 22. Anchorage was dropped and on the third vote Lillehammer received 45 votes (54%, a majority!) to win. If it strikes you as odd that Ostersund went from third to first to second, you might wonder if this is a good voting system.

While "majority rules" may seem like the obvious approach to any election, we are about to see why we need alternatives.

Condorcet's Intransitive Attitude

The transitive property is an example of an idea that we often take for granted, but which we assume to be true at our own peril. The real numbers have an ordering, so that we can always rank them, with the **transitive property** that if $a < b$ and $b < c$, then $a < c$. This means that once we know that $2 < 3$ and $3 < 5$, we can immediately jump to the conclusion that $2 < 5$. In sports, if we believe that Dallas is better than New York and New York is better than Washington, then we surely believe that Dallas is better than Washington. This belief is necessary if we are going to rank teams.

Trivial and boring? Try this. Early in the season, Dallas has defeated New York and New York has defeated Washington. Does this guarantee that Dallas will defeat Washington? Of course not. Upsets happen, teams have bad days, balls bounce funny, and referees make calls. Sports results are *not* transitive. This, in fact, is part of the fun of following sports: the possibility of a surprising, illogical result is always there.

Transitivity is a basic assumption underlying ranking systems. When we list our top 5, we mean that number 1 is better than number 2, and number 1 is better than number 3, and so on. However, we recognize that sports *results* are not transitive. Voting, it turns out, is decidedly *not* transitive.

Suppose that a three-person committee needs to rank three teams A, B, and C. One person ranks the teams in the order A-B-C, the second person ranks them in the order B-C-A, and the last person ranks them in the order C-A-B. Each team is tied with one first-place vote. Believing in the majority-rules principle, we look at two teams at a time. Given a choice between only A and B, two out of three committee members (the first and third) prefer A over B. Also, two out of three committee members prefer B over C. The committee prefers A over B, and prefers B over C; do we even need to check out A versus C? Actually, we do, because two out of three committee members prefer C over A!

In voting situations with three or more candidates, the transitive law *does not hold* in the sense that the majority of voters can prefer candidate A over B, B over C, and C over A. The problem is not one of quirky personalities or voters changing minds or being illogical. The transitive law is simply not true for voting situations of this type.

FIGURE 6.1: The Author and Condorcet in Paris

The intransitivity of pairwise voting has been known for hundreds of years, going back to the first analytical study of voting. In 1785, the Marquis de Condorcet published a theory of voting, a logical outgrowth of the rational idealism spawned by the American Revolution and its influence in France. Condorcet proposed the two-at-a-time process described above, and stated that the candidate who wins all head-to-head matchups should be the overall winner. The problem, as Condorcet himself discovered, is that there is not always such a winner. In our three-team example above, no team wins all of its head-to-head battles; we say that A, B, and C form a **Condorcet cycle**.

For Condorcet, the search for a perfect voting system hit a dead end. For us, the lack of a transitive property for group voting means, for starters, that voting methods are different from ranking systems.

Preference Lists, Voting Systems, and Chaos

We look at situations where a group of voters is trying to rank some number m of candidates with $m > 2$. Each voter has a **preference list** of candidates; i.e., a ranking of candidates. The preference list for each voter obeys the transitive property; an individual who prefers A over B and B over C also prefers A over C. Based on this collection of preference lists, we want to identify the winner of the election.

We have already defined the **plurality method**. The candidate who receives the most first-place votes wins the election; we call this candidate the **plurality winner**. The NFL MVP is decided by the plurality method. In this method, as with the others to follow, there is the possibility that a tie will need to be broken. For simplicity, we make the assumption that ties are broken in some logical manner (which we will not address specifically).

The **Condorcet winner** of an election is a candidate, *if one exists*, who wins every head-to-head matchup with other candidates. The problem, as we have seen, is that in some cases there is no Condorcet winner, due to a Condorcet cycle. Of course, a necessary property for a useful voting method is that it determines a winner in every election, so this is a big problem. This is also an issue for the **majority method**, in which a candidate receiving a majority (more than 50%) of the votes is the winner.

A popular voting method is the **Borda count**. Each candidate receives $m-1$ points for each first-place vote, $m-2$ points for each second-place vote, and so on down to 0 points for each last-place vote. The **Borda winner** is the candidate with the most total points. The Heisman Trophy, the MLB MVP and the NBA MVP are versions of the Borda count. Voters for these awards are not asked to rank *all* candidates, however, and the points awarded vary. The Borda count was invented by Condorcet's contemporary and rival, Jean Charles de Borda, and later re-invented by Charles Dodgson (aka Lewis Carroll of *Alice's Adventures in Wonderland* fame).

The Olympic Games method of selecting host cities is an example of a **plurality with elimination** method. In this method, successive votes are taken until a candidate receives a majority of votes; this candidate is declared the winner. If a vote does not result in a majority decision, the candidate with the fewest votes is eliminated and a new vote is tabulated. For the Olympics, separate votes are taken, allowing voters to change their minds. In **instant runoff voting**, each voter submits a preference list and all rounds of voting are conducted from this list. For example, a voter with preference list A-B-C-D would cast a vote for A; if A is eliminated, that voter's vote would go to B. If B is also eliminated, the vote next goes to C, and so on.

The **approval voting** method allows voters to vote for any and all candidates of which they approve. That is, for each candidate the voter says "yes" or "no" and the candidate with the most yes-votes is the **approval win-**

ner. The Baseball Hall of Fame voting is similar to approval voting, although each HOF voter is limited to no more than ten votes. Further, all candidates who receive more than 75% of the possible votes are elected, so there can be multiple winners.

Example 6.1

4	6	9	11
A	A	B	C
B	C	A	B
C	B	C	A

For the preference lists given, find the following (if it exists): (a) majority winner, (b) plurality winner, (c) Condorcet winner, (d) Borda winner, (e) plurality with elimination winner.

Solution. The table indicates that 4 voters rank the candidates in the order A-B-C, 6 voters rank the candidates in the order A-C-B, and so on. So 4+6=10 voters cast first-place votes for A, 9 cast first-place votes for B, and 11 cast first-place votes for C.

(a) With 30 voters, a majority is at least 16 votes, so there is not a majority winner. (b) C has the most first-place votes (11) so C is the plurality winner. (c) Matching A and B head-to-head, the 4 voters in the first column and the 6 voters in the second column prefer A over B, while the remaining 9+11 = 20 voters prefer B over A. B beats A, 20-10. Matching B and C, the 4 voters in the first column and the 9 voters in the third column prefer B over C, but the other 6+11 voters prefer C; C beats B, 17-13. Matching A and C, A gets the voters in the first three columns and wins 19-11. Each candidate wins one and loses one head-to-head matchup, so there is no Condorcet winner.

(d) For the Borda count, note that A gets 4+6=10 first-place votes, 9 second-place votes, and 11 third-place votes. Then compute

A : 2x10 + 1x9 + 0x11 = 29 points

B : 2x9 + 1x(4+11) + 0x6 = 33 points

C : 2x11 + 1x6 + 0x(4+9) = 28 points

so B is the Borda winner. (e) For plurality with elimination, start with first-place votes of 10 (A), 9 (B), and 11 (C). Nobody has a majority, so we drop the lowest vote-getter, which is B. With B eliminated, A now gets the 9 votes from B's supporters and beats C 19-11. A wins plurality with elimination.

Example 6.1 should be disturbing. Depending on the voting method used, any of the three candidates can win! Of course, this example was specially constructed to work this way. In many elections, there is a clear-cut winner and the voting method used does not matter. In close elections with three or more candidates, however, you should be aware that the winner might depend on the voting method used.

Fairness and the Arrow of Impossibility

In a close election, different voting methods can produce different results. This leads us to investigate which method is the best and which methods should never be used. Taking this negative angle, we want to reject any method that behaves in an illogical manner. The **fairness criteria** make precise some of the ways that a voting method could misbehave.

We want voting methods to satisfy **monotonicity**: if candidate A would win using a set of preference lists and the only changes that are made to the preference lists move A *higher* in the list(s), then A should still win. Think back to Example 6.1(e). With the voter preference lists as given, A wins using plurality with elimination. Now, suppose that right before the vote occurs A makes an announcement that sways some of the voters in A's favor (no jokes about the Qatar World Cup selection, please). Before looking at the details, let's summarize: A was going to win the election anyway, and then gains more support. Logically, A should still win, right? This is what we mean by monotonicity.

Split the fourth column of 11 voters in Example 6.1 into 8 who maintain the same order and 3 who move A to the top (making their preference A-C-B). Now A has 13 first-place votes, B has 9 votes and C has 8. We eliminate C, and then match A and B head-to-head. B gets 9+8 votes and beats A, 17-13, to win the election. What happened? Three voters improved their opinion of A and this cost A the election! Technically, we say that plurality with elimination violates the monotonicity fairness criterion. More informally, we can cross that method off the list of good methods.

Another important fairness criterion goes by the unwieldy name of **independence of irrelevant alternatives (IIA)**: if a set of preference lists is changed in a way that leaves the relative ordering of candidates A and B the same, then the voting method should not change its ordering of A and B. In other words, once the voting populace has decided that it prefers A to B, additions, deletions, or re-orderings of other (irrelevant) candidates should not reverse the decision.

A famous violation of IIA occurred at the 1997 European Men's Figure Skating Championships. At the time, the system used to rank competitors was complicated (see exercise 6.19), but involved two rounds of skating. With one skater left in the final round, the official standings were Vyacheslav Zagorodniuk in first, Alexei Urmanov second, Ilya Kulik third, and Philippe Candeloro fourth. The last skater ended up finishing sixth, so logically the top four should remain the top four, right? However, the order changed completely: Urmanov moved up to first, Candeloro was second, Zagorodniuk third, and Kulik fourth. Zagorodniuk fell from first to third without skating, and with only the sixth-place skater doing anything! This final result was definitely dependent on the irrelevant alternative of the last, sixth-place skater.

Unfortunately, both plurality and the Borda count can violate IIA. In Example 6.1(b), C wins plurality over A and B, but if in the third column A and B are reversed (which should be irrelevant since both are already ranked above C in this list) then C loses and A wins. In Example 6.1(d), B wins the Borda count, but if A and C are flipped in the second column then C wins.

We now have reasons to throw out all five of our voting methods: majority and Condorcet winner do not always produce winners, and the other three violate fairness criteria. The hope of finding a perfect voting method was skewered in 1948 by Kenneth Arrow, who won a Nobel Prize for this and other work. **Arrow's Impossibility Theorem** states that among voting methods that rationally use voter preference lists to rank three or more candidates, the only method that does not violate at least one of the two fairness criteria (monotonicity and IIA) is a dictatorship (that is, a situation where a single voter makes all decisions). In short, there is no perfect voting method based on preference lists. Note that approval voting does not use full preference lists (if a voter approves of candidates A and B, we cannot tell whether or not A is preferred over B) and so is not judged by Arrow's Impossibility Theorem. As we will see, approval voting has its issues.

There are further properties we might hope that a voting method would satisfy. For example, when a majority winner or Condorcet winner exists, we want our voting method to identify that candidate as the winner. Approval voting violates both of these criteria (as does the Borda count; plurality violates the Condorcet criterion). Each voting method has its own set of eccentric behaviors that make it appear unfit for usage.

As discouraging as that may seem, mathematically speaking it just means that the research continues. If there is no perfect method, there could still be a *best* method, one which is least likely to violate fairness criteria or one which is optimal by some other measure. The search continues in the next section.

Positional Voting Systems

Mathematical results have been proved for voting systems that can be framed as **positional voting systems**. The concept is easy to understand with the Borda count. In Example 6.1, there are three candidates and the Borda count assigns 2, 1, and 0 points to votes. We can represent this method as a vector [2, 1, 0]. If we want to emphasize first-place votes by assigning them 5 points, our system is represented by [5, 1, 0]. If we want to use plurality, the system is described by [1, 0, 0], since only a first-place vote counts. Approval voting would be described by [1, 1, 0] if all voters approved of exactly two candidates, but since this is not always the case approval voting is not a positional voting system.

Positional voting systems include three of the voting methods discussed above. Plurality with elimination and approval voting do not fit this system.

Mathematicians have done extensive work on positional voting systems, and have proved a result that is suggestive for sports voting. The result relates to positional voting systems with *full* ranking of candidates (each preference list ranks all candidates). Sports situations where a voter submits a top ten or top twenty are not full rankings, since most teams are not included in the lists.

However, in the case of full rankings the Borda count is the positional voting system that maximizes the likelihood that the Condorcet winner is ranked first. Thus, changing a [2, 1, 0] Borda count to a [5, 1, 0] system with a bonus for first place votes makes violations of the Condorcet winner criterion *more* likely.

This does not prove that the unequal weights for MVP votes are suboptimal, since MVP votes are not based on full rankings. However, it does suggest that excessive tinkering with the points for first place could be counterproductive.

A Return to Sports Voting

With some mathematical terminology and results in hand, let's return to our sports examples for some evaluations.

The Heisman Trophy uses a (partial ranking) Borda count [3, 2, 1]. There are problems with the Borda count. As Tim Tebow found out, the plurality winner also has no guarantee of being the Borda winner; even a majority winner can lose the Borda count. The Borda count also violates IIA.

On the plus side, the Borda count allows voters to give support to a variety of candidates. Further, suppose that there are two top candidates A and B for the Heisman. You favor A. In one case, you find it difficult to distinguish between A and B; in another case, you think that B is unworthy of the award. With plurality, you vote for A. With the Borda count, you can give B second-place points or worse, so your opinion of B counts.

The NFL MVP is a plurality vote. In close elections, the plurality system is unusually prone to violating head-to-head preferences of the voters, as we see in the simulation results that follow. The main positives of the plurality system are its ease of implementation, the lack of knowledge required of voters (you only need to pick out one candidate; perhaps for an important award this should not be considered a positive), and its familiarity.

Baseball and basketball MVPs are decided by (partial ranking) Borda counts with extra points for first place votes. We see in Figures 6.2 and 6.3 that the extra points make little difference in the ability of the system to agree with head-to-head preferences. Philosophically, this type of system is a compromise between the broad-based support of a Borda count and the familiarity of the plurality system. Mathematically, it can be viewed as a weighted sum of the Borda count and plurality systems, with the plurality component polluting the better properties of the Borda count.

The Olympics selection process uses plurality with elimination. This system violates IIA and monotonicity, with the monotonicity issue making the process seem almost random for close elections. On the plus side, a Condorcet winner (if one exists) will always be elected and elimination eventually boils down to a majority-rules decision.

The baseball HOF uses a modified approval voting method, with each voter limited to ten approvals and all candidates with at least 75% of the votes elected. Approval voting does not directly violate Arrow's Impossibility Theorem, but the HOF modification does violate IIA (see exercise 6.26). The method used allows for unequal numbers of candidates to be elected in different years, giving the process a feeling of a knowledgable thumbs up or thumbs down on each candidate.

Simulations

Other than the Olympics selection committee, sports voters do not use a pure version of any of the basic voting methods. While this is done for sound practical reasons (we should not really care whether a voter ranks Sammy Watkins 23rd or 24th in MVP voting), it does mean that the mathematical beauty of Arrow's Impossibility Theorem does not directly apply. If there are no theoretical results to evaluate our voting methods, we can still collect evidence from simulations. Because most simulations include some element(s) of randomness, simulation results are rarely conclusive. Nonetheless, seeing the voting methods in action can give us important information.

The first round of simulations applies to 50 players who have a "true" ability. The abilities are not bunched: the best player has a value of 50, the second best 49, and so on. The voters have a fuzzy evaluation of the players: each voter rates each player as $v + e$ where v is the true value of the player, and e is a random error in judgment. So, a given voter might rate the first player as $50 - 3.4 = 46.6$ and the second player as $49 + 0.3 = 49.3$ and conclude that the second player is better. The simulation takes these flawed ratings and orders them into preference lists for 60 voters. The preference lists are then run through three voting methods: plurality, a [20, 19, 18, ..., 2, 1] partial ranking Borda count for the top twenty, a [25, 19, 18, ..., 2, 1] count that gives 5 bonus points for first-place votes, and approval voting (scores above 45 - on average, the top five - are approved). This is done 1000 times and the results compiled.

To see which voting methods follow the "will of the people," we look at how often the voting methods agree with the opinions of the voters (whether they were "right" or not). If a majority of voters prefer A to B, we check whether the voting method ranks A over B. If it does, the voting method "wins" that matchup. For all four methods, there are "ties" where both A and B receive 0 points. This is especially true for the plurality method, since the 50th best player is unlikely to receive a first-place vote. Ties are ignored, and we compute a winning percentage for the cases where the voting method has an opinion.

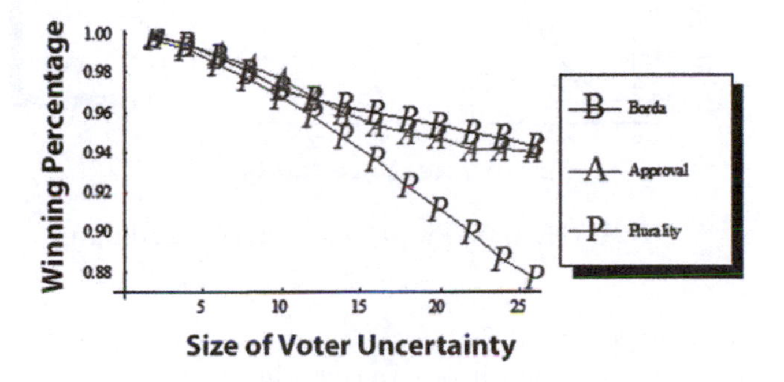

FIGURE 6.2: Borda Count, Approval, Plurality

Figure 6.2 shows the results for a variety of average voter evaluation errors. The higher the error, the less certain the voters are about the true quality of the players, and therefore the more random their votes are. Only three curves are distinguishable, because the winning percentages for the two Borda counts agreed to three decimal places. For small errors, the ordering of players is clear to most voters, and all voting methods perform well. When there is substantial disagreement among the voters as to who is best, plurality does a noticeably worse job of matching voter sentiment than the Borda count and approval voting.

The results change little when the first-place bonus points increase to 20. When the threshold for approval voting is lowered from 45 to 40, the method does better for large voter evaluation errors, essentially matching the Borda count winning percentages.When the number of voters is increased to 300, the winning percentages of all three methods increase.

A second round of simulations explores what happens when two controversial candidates are in the pool. The controversy may be over the meaning of "valuable" (e.g., a player on a losing team) or over some social issue (e.g., a steroid user). For each controversial candidate, about one-third of the voters start with a base value that puts the candidate in first place. The other two-thirds have a starting value in tenth place (losing team) and twentieth (steroids).

Plurality is clearly inferior here. Although its winning percentage increases as voter errors increase, this is not a ringing endorsement for a voting method.

("It's better when nobody knows what they're voting on!") You can see the effect of giving bonus points for first place in this simulation, with the bonus Borda distinctly below the regular Borda count.

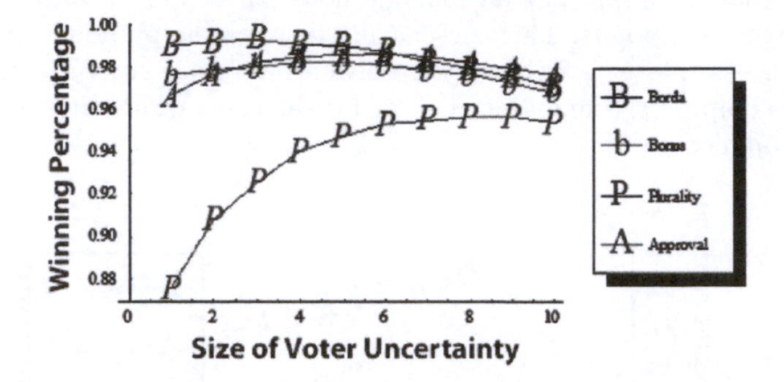

FIGURE 6.3: Two Controversial Candidates

It is important to repeat that simulations do not prove anything. However, plurality does not look good here. We sometimes unthinkingly act like this is the only way to vote, but most voting methods experts consider plurality one of the worst voting methods available. By association, the bonus Borda also looks inferior. The voting methods research community has a lively debate over whether the Borda count or approval voting is better. The simulations give an edge to the Borda count if most voters only approve of a small number of candidates, with the two methods tied for more generous approval voters.

Range Voting

The old figure skating voting system gave us an egregious violation of IIA, when the leaderboard at a major competition was suddenly scrambled. To avoid such embarrassments, the scoring system was completely revised. The new system is a point system, where nine judges give the skaters grades, the highest and lowest are discarded, and the average of the seven remaining scores is retained. Similar systems are used in other sports, including gymnastics and diving.

This is called a **trimmed mean**, and the voting system is similar to a system called **range voting**. We use range voting regularly to provide feedback on purchases at such places as Amazon. Range voting is not based on preference lists. Voters assign each candidate a rating (e.g., five stars) and may give the same rating to many candidates. Because the outcome is not based on preference lists, Arrow's Theorem does not apply to range voting. In fact, range voting does not violate monotonicity (moving a candidate up gives the candidate more points, which can only help) and does not violate IIA (the points assigned to two candidates do not change if other candidates are added, deleted, or modified).

Even though it passes the criteria of Arrow's Theorem, range voting can violate a majority decision if the minority ranks the candidate low enough. This indicates that range voting is susceptible to insincere or strategic voting. Suppose you want player A to win the MVP, and it is clear that player B is A's main competition. Even though you respect player B, you might decide to improve A's chances by giving A the maximum vote and B the minimum vote. If you also give other candidates the minimum score, what you have done is equivalent to plurality voting, giving one candidate one vote. Since plurality does not rate highly as a voting method, strategic voting could reduce range voting to an inferior product.

You might expect that figure skating, which has a history of manipulated votes, is subject to this flaw. However, it seems that reviews of judges' records has been enough to keep most judges honest.

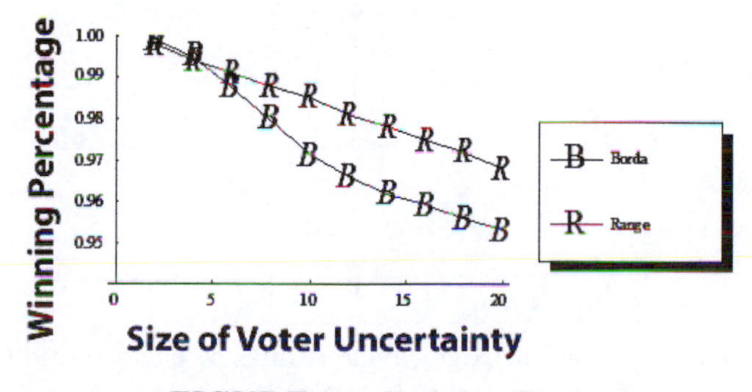

FIGURE 6.4: Range vs Borda

The results from adding range voting to the simulation of Figure 6.2 are shown in Figure 6.4. To simulate range voting, each voter's candidate ratings is converted to an integer from 1 to 5. (Using the actual voter rating is more accurate, but it would be unrealistic to ask real voters to provide ratings with 3 significant digits.) As it is, range voting is better than the Borda count, and hence better than approval voting and plurality, at identifying head-to-head voter preferences.

A sophisticated simulation published in *Gaming the Vote* took into account a variety of possibilities, including insincere voting and voter ignorance of some candidates. In this case, the criterion for best voting method was to maximize the overall "happiness" of the voting populace. Range voting was clearly better than approval voting, Borda count, and the other methods discussed here. Of these, plurality came in last.

PageRank and MVPassing

A search engine gives you an ordered list of results. You expect that the first results listed will be the most useful, but how does the search engine decide which results to present first? In this section, we discuss the original formulation of Google's PageRank for determining the importance of a web page. The importance of the page is combined with the relevance of the page to the keywords in the search to produce the order of the search results.

A voting analogy help us understand how PageRank works. View web pages as shareholders. Suppose that page A has 10 shares of "stock" and links to five other pages. Then A is voting 2 votes apiece to those 5 pages. In turn, A gets its votes when other pages link to A. The more links A has, and the more shares each of those pages has, the higher A will rank.

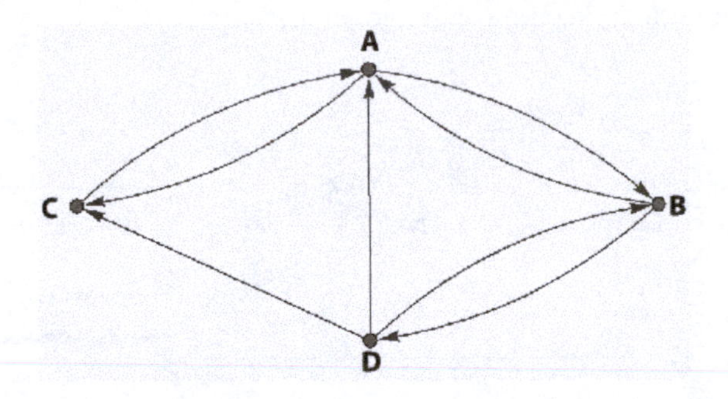

Imagine a "web" of four pages. Page A links to pages B and C; if A has a shares, then B and C each receive $\frac{1}{2}a$ votes. If page B links to A and D, each receives $\frac{1}{2}b$ votes. If page C only links to A, then A receives c votes. If page D links to all three pages, each receives $\frac{1}{3}d$ votes. In all, page A has received $\frac{1}{2}b + c + \frac{1}{3}d$ votes, so $a = \frac{1}{2}b + c + \frac{1}{3}d$, and we have similar equations for b, c, and d.

Does this look familiar? It should remind you of equations for the Massey and Colley systems in Chapter 5. PageRank can be thought of as a matrix-based rating system. Another view of PageRank is developed in exercise 6.23.

Example 6.2 For the four-page web described above, find PageRank values that sum to 1.
Solution. The equations for the ratings a, b, c, and d are given by
$a = \frac{1}{2}b + c + \frac{1}{3}d$
$b = \frac{1}{2}a + \frac{1}{3}d$
$c = \frac{1}{2}a + \frac{1}{3}d$
$d = \frac{1}{2}b$
which can be summarized in the matrix

$$\begin{bmatrix} -1 & 1/2 & 1 & 1/3 & 0 \\ 1/2 & -1 & 0 & 1/3 & 0 \\ 1/2 & 0 & -1 & 1/3 & 0 \\ 0 & 1/2 & 0 & -1 & 0 \end{bmatrix}.$$

You can verify that this system of equations has an infinity of solutions. This is why we add the requirement that the ratings sum to 1; that is, we add the equation $a+b+c+d=1$ to get the matrix

$$\begin{bmatrix} -1 & 1/2 & 1 & 1/3 & 0 \\ 1/2 & -1 & 0 & 1/3 & 0 \\ 1/2 & 0 & -1 & 1/3 & 0 \\ 0 & 1/2 & 0 & -1 & 0 \\ 1 & 1 & 1 & 1 & 1 \end{bmatrix};$$

and then reduce this matrix to get

$$\begin{bmatrix} 1 & 0 & 0 & 0 & 10/25 \\ 0 & 1 & 0 & 0 & 6/25 \\ 0 & 0 & 1 & 0 & 6/25 \\ 0 & 0 & 0 & 1 & 3/25 \\ 0 & 0 & 0 & 0 & 0 \end{bmatrix};$$

so $a = 10/25$, $b = c = 6/25$, and $d = 3/25$. A is the most important page, B and C are tied for second, and D is last. If you think through the linking structure, this should make sense. All pages link to A, and only B links to D. B and C are both linked to by A and D.

There are interesting applications of PageRank to sports. Using PageRank to rank teams, as in Chapter 5, is not one of them. For web pages, a page votes for another page if it links to that page. For teams, we could say that a team votes for another team if it loses to that team. The issue of what to do with an undefeated team is explored in exercise 6.23. A larger problem occurs with a team that loses only once. This is a very good team and should be highly rated. The team that beats it gets *all* of its voting shares. In the case of a large upset, this gives the underdog team far too much credit.

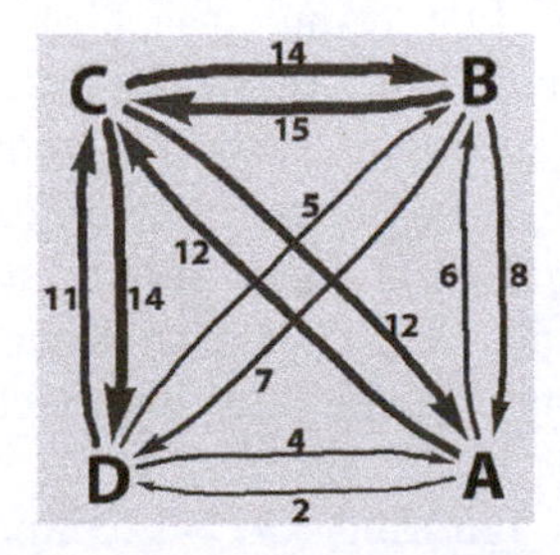

PageRank has been applied with good effect to passing networks in sports such as soccer and field hockey. Soccer teams have different passing tendencies. Barcelona is known for incessant short passes, while other teams are known for long balls. We can count the number of passes made between players, and this may give us information about tendencies. The data may show that a team favors the left side, but that is likely to be known to the players. A harder task might be to identify the most important link in the passing structure (the MVPasser), analogous to the most important page on the web.

Suppose our team has four players. During the course of the game, player A passes to player B 6 times, player C 12 times, and player D 2 times. This is shown in the graph above, along with the remainder of the passing data. The arrows of the graph (we call them *directed edges*) are sized according to how many passes are made, so we can see quickly that A and D do not interact much.

The matrix for the ratings a, b, c, and d is given by

$$\begin{bmatrix} -1 & 8/30 & 12/40 & 4/20 & 0 \\ 6/20 & -1 & 14/40 & 5/20 & 0 \\ 12/20 & 15/30 & -1 & 11/20 & 0 \\ 2/20 & 7/30 & 14/40 & -1 & 0 \\ 1 & 1 & 1 & 1 & 1 \end{bmatrix}.$$

The first column shows that 6 out of A's 20 passes are directed to B, 12 out of 20 to C, and 2 out of 20 to D. Reducing the matrix gives us the ratings $a = .209$, $b = .237$, $c = .354$, and $d = .200$. This tells us that player C is the MVPasser. In this case, the ratings have direct meaning. Over the course of an infinitely long game, C has control of the ball 35.4% of the time, B has the ball 23.7% of the time, A 20.9%, and D 20%. The percentages are not measured in clock time; a better way of phrasing the result is that 35.4% of the passes go to player C.

Seeding of Tournaments

The seeding of tournaments is designed to separate the best players or teams so that they do not face each other early in the tournament. In the NCAA basketball tournament, all teams are seeded. Do the seedings matter? In the spirit of our voting method analysis, we want to ask whether any paradoxes occur in which it is better to have a worse seed.

At first glance, the answer appears to be "no." Suppose that the tournament has 64 teams in four regions. In each region, teams are ranked 1-16. In the first round, the best team (seed 1) plays the worst team (seed 16), the second-best team (seed 2) plays the second-worst team (seed 15), and so on. Figure 6.5 shows the frequency with which a given seed won its first round game in the years 1979-2014. With the exception of seeds 5 and 12, the percentage is decreasing in an orderly linear fashion. Seeds 5 and 12 play each other, and it is a well-known part of "March madness" that the 12-seeds always provide an upset victory or three. In the first round, the better the seed the better the result.

Ignoring the 5-12 blip, we can fit a line $y = 1.0627 - .0662x$ to the data for the probability that seed x wins its first-round game. Our question about seeding is not fully answered, however. In the second round games, we expect to have seeds 1 and 8, 2 and 7, 3 and 6, and 4 and 5 play. However, suppose that both the 9-seed and 10-seed win in the first round. Then 1 plays 9 and 2 plays 10; the 10-seed has an easier game! To see how these matchups play out over the course of the tournament, we can use simulations. To do so, we need probabilities for teams winning games in each round.

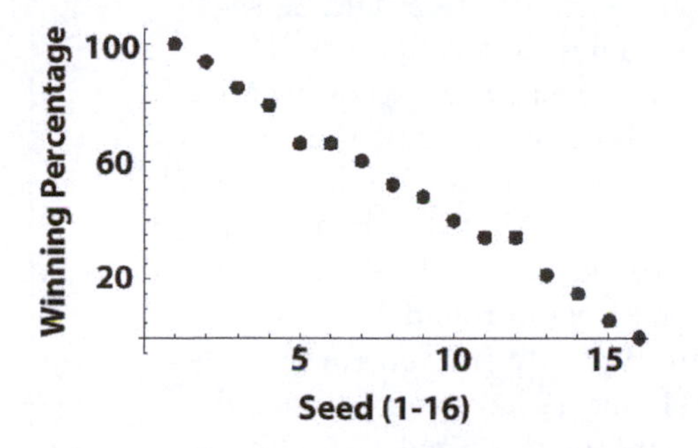

FIGURE 6.5: First Round Wins by Seed

The expected second-round games (1-8, 2-7, 3-6, 4-5) follow a similar trend to Figure 6.5. The exception is the 6-seed beating the 3-seed more often than expected. The line $y = .896 - .083x$ is a good fit. For example, 2-seeds beat 7-seeds with probability .730 while 1-seeds beat 8-seeds with probability .813. If the 10-seed upsets the 7-seed in the first round, a good approximation for the 2-10 second round game is to boost the 2-seed's probability one level, from .730 to .813. The logic is that 1-8 and 2-10 games are similar in the degree of mismatch. Implicit in the logic is a re-evaluation of the quality of the teams. A 15-seed who has reached the third round has proven to be underrated, and so is given a better chance to win. Figure 6.6 shows the results of a million simulated tournaments.

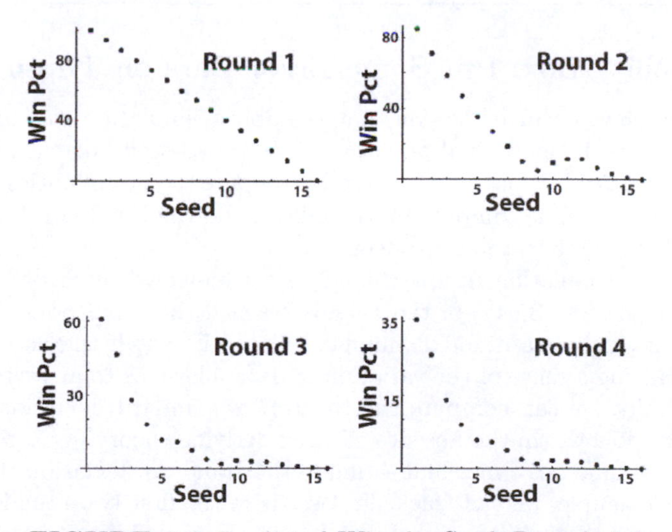

FIGURE 6.6: Predicted Wins by Seed, Rounds 1-4

Beyond round 1, it is *not* true that better seeds get better results. Figure 6.6 shows that more 10-seeds, 11-seeds, 12-seeds, and 13-seeds win second round games than do 9-seeds. This is because 9-seeds have to play the 1-seeds, whereas 10-seeds play 2-seeds or, sometimes, 15-seeds. In the third round, 6-seeds win as often as 5-seeds. Again, the explanation is that it is bad news to play a 1-seed. In round 3, 5-seeds almost surely play the 1-seed. Winners in round 4 (the winner of the region) are typically the top three seeds, but seeds 8, 10, 11, and 12 have nearly equal likelihoods of advancing to the Final Four. To the right is our reality check: the actual breakdown of regional winners for 1979-2014 next to the predictions from the simulations. The biggest tournament surprises include the three 11-seeds (VCU, George Mason, and LSU) who reached the Final Four. The simulation predicts three 11- or 12-seeds out of the 144 regions in 36 years, and that is exactly what has happened! Yes, lumping seeds together is cheating, but the predictions are reasonably accurate. Of importance for us is the conclusion that 10-, 11- and 12-seeds all have a better chance of reaching the Final Four than 9-seeds.

Seed	Actual	Pred.
1	39.6	35.4
2	21.5	24.1
3	11.1	13.2
4	9.7	8.3
5	4.8	5.4
6	4.2	4.4
7	1.4	3.0
8	4.2	1.5
9	1.4	0.7
10	0	1.1
11	2.1	1.1
12	0	0.9

FIGURE 6.7: Percentage in Final Four by Seed

Probability Box: Put Some Error Bars on Those Things

It may have occurred to you that the above simulation was unnecessary. Once we have the individual probabilities (e.g., a 3-seed beats a 15-seed with probability 0.79) it is possible to directly compute the probabilities of reaching different rounds. The computed probabilities and the Final Four probabilities shown above agree to the third decimal.

However, simulation results should be accompanied by error bars. In the NCAA simulation, 35.4% of the 1-seeds reached the Final Four. Will a new simulation produce a different number? Could it match the actual value of 39.6%? Having a sense of the variation in data allows us to answer such questions. Ideally, we can compute the theoretical standard deviation; a sample calculation follows. In the absence of an underlying theory, we can repeat the simulation numerous times and estimate the standard deviation that way.

As an example, we can track the 1-seeds in the first two rounds. In round one, the 1-seeds win with probability 0.9965. In probability terms, we model the games as *Bernoulli trials* with $p = 0.9965$; in each game, the 1-seed wins with this probability. For a simulation of one million trials (n), the standard deviation of the proportion of wins equals $\sigma = \sqrt{\frac{p(1-p)}{n}} = 0.000059$. Less than 5% of the simulations should produce a winning proportion more than 2σ away from p, outside of the range $0.9965 \pm .000118$. So, the simulated first round should be very accurate.

To make it through two rounds, the 1-seed must win in the first round and then beat either the 8-seed or the 9-seed. We multiply probabilities to get a combined probability of $0.9965\, p_2$ for a second round win, where p_2 is the probability that the 1-seed wins its second round game. There are two possibilities: the 1-seed beats the 8-seed or the 9-seed (whichever won in the first round). We need the probability that #1 beats #8 (0.813) in round two times the probability that #8 beat #9 in round one (0.5331), plus the probability that #1 beats #9 (0.896) times the probability that #9 beat #8 (0.4669). The probability that a 1-seed wins in round two is $0.9965(0.813 \cdot 0.5331 + 0.896 \cdot 0.4669) = 0.8488$.

The calculation of the standard deviation gets messy. With the notation $p_1 = 0.9965$ and $\sigma_1 = 0.000059$ for the first round mean and standard deviation, respectively, and p_2 and σ_2 for the corresponding second round values, the standard deviation for second round wins is $\sqrt{p_1^2\sigma_2^2 + p_2^2\sigma_1^2 + \sigma_1^2\sigma_2^2}$. We have $p_2 = 0.813 \cdot 0.5331 + 0.896 \cdot 0.4669 = 0.852$ and can compute $\sigma_2 = 0.000654$. The second round standard deviation is 0.00056. About two-thirds of the simulations should produce a 1-seed second-round probability in the range 0.8488 ± 0.00056. Our simulation value of 0.8484 is in the lower end of that region.

The bottom line is that the error bars for Figure 6.6 would be invisible, because the standard deviations are quite small for simulations of one million tournaments.

Exercises

In these exercises, Ⓣ refers to thinking problems, conceptual problems requiring no calculations. Ⓒ refers to problems requiring significant proof-writing ability. Ⓟ refers to projects; these are ideas for further investigation (hints and resources are at the book's web site).

6.1 (a) Suppose that in the 2008 Heisman voting points were only given for first and second place votes, with 3 points for a first place vote and 1 point for a second place vote. Would the results have changed? (b) Assigning x points for first place votes, 2 for second, and 1 for third, how large would x have to have been for Tebow to win the 2008 Heisman? (c) If the Heisman voting were changed to a HOF-like system where voters vote for three players, who would have won the 2008 Heisman? (d) Are there any years in which this system would have changed the winner?

6.2 Determine whether the Heisman voting procedure obeys majority rules, meaning that a player who receives a majority of first place votes always wins.

6.3 For the 2005 NFL MVP voting, Shaun Alexander received 19 of the 50 votes, followed by Peyton Manning with 13 and Tom Brady with 10. If the voting system used was to receive 2 points for a first place vote and 1 point for a second place vote, give a hypothetical circumstance in which the winner could have been (a) Manning, (b) Brady.

6.4 In the 1999 American League MVP voting, Pudge Rodriguez received 7 first place votes and 252 points to edge Pedro Martinez, who received 8 first place votes and 239 points. If both players received 2 7th-place votes, 2 6th-place votes, 2 5th-place votes, 3 4th-place votes, and 5 3rd place votes (this is hypothetical), how many 2nd-place votes did each player receive? In this situation, how many of the 28 voters left Martinez off their ballots? Why might a voter do that?

6.5 Explain how you know that at least one voter changed his or her preference list in the 1994 Olympics vote. Recall that the votes were taken at different times. Describe a scenario in which it would be reasonable to change your mind.

6.6 Suppose that in voting for the 1994 Olympics site, 25 voters had the preference list L/A/O/S, 17 voters had the preference list A/S/L/O, 6 voters had the preference list A/O/S/L, 19 voters had the preference list O/A/L/S, 14 voters had the preference list S/O/A/L, and 3 voters had the preference list S/L/A/O. (These totals are made up.) Determine the winner using plurality with elimination. If a third- or fourth-place vote represents "disapprove" which city has the (a) highest (b) lowest percentage of "disapprove" votes?

6.7 In the scenario of exercise 6.6, suppose that at each stage the city with the most last-place votes is eliminated. Show that Anchorage would be chosen. Discuss which elimination rule (most last-place votes or fewest first-place votes) makes more sense.

6.8 In an election with three candidates, 8 voters have the preference list A/B/C, 5 voters have B/C/A, 3 voters have C/A/B, and 3 voters have C/B/A. Find the winner using (a) plurality; (b) Borda count; (c) plurality with elimination. (d) Is there a Condorcet winner? (e) If each voter approves of his or her first two candidates, who wins approval voting?

6.9 In an election with four candidates, 7 voters have the preference list A/D/B/C, 5 voters have C/A/B/D, 4 voters have C/B/D/A, and 3 voters have D/A/B/C. Find the winner using (a) plurality; (b) Borda count. (c) Is there a Condorcet winner? (d) A **Condorcet loser** is a candidate who loses every head-to-head matchup. Show that the plurality winner is a Condorcet loser. Comment on why many feel that plurality is a poor voting method.

6.10 In an election with four candidates, 10 voters have the preference list A/B/C/D, 8 voters have C/B/A/D, 6 voters have C/D/B/A, and 5 voters have B/C/D/A. Find the winner using (a) plurality; (b) plurality with elimination; (c) Borda count. (d) Find the Condorcet winner. (e) Find the Borda count winner if D drops out, leaving three candidates. Which fairness criterion has been violated?

6.11 For each method, give an example showing that it violates the Condorcet criterion: (a) plurality; (b) Borda count; (c) approval voting; (d) range voting.

6.12 Suppose teams are rated by some method and every game conforms exactly to the ratings (i.e., if team A is rated 5 points better than B, then A beats B by 5 points). Explain why a Condorcet cycle is not possible. In other words, Condorcet cycles are caused by deviations from teams' ratings.

6.13 Verify that (a) if A and B are reversed in the third column of example 6.1(b), C loses plurality; (b) if A and C are reversed in the second column of example 6.1(b), B loses the Borda count. (c) Explain the significance of parts (a) and (b) in terms of the fairness criteria.

6.14 The Borda count can be used to rank all candidates; in Example 6.1, the ranking would be B (33 points), then A (29 points), then C (28 points). Explain how to use (a) plurality and (b) plurality with elimination to rank all candidates. (c) Suppose you have these rankings (no ties). If all voters reverse their preference lists (e.g., change A/B/C to C/B/A), will the rankings be reversed? Determine yes or no for each of the three voting systems, and comment on whether you think a good ranking system should reverse the order.

6.15 Show that the positional voting systems [4, 3, 2, 1] and [3, 2, 1, 0] are **equivalent**, meaning that they will always produce the same rankings as each other. Show that the positional voting systems [6, 4, 2, 0] and [3, 2, 1, 0] are equivalent. Show that [6, 3, 1] and [3, 2, 1] are not equivalent by finding a set of preference lists in which the systems produce different winners.

6.16 Determine whether the positional voting systems are equivalent or not. (a) [4, 3, 2, 1] and [7, 5, 3, 1]; (b) [4, 3, 2, 1] and [10, 7, 4, 1]; (c) [4, 3, 2, 1] and [8, 4, 2, 1]; (d) [8, 4, 2, 1] and [27, 9, 3, 1].

6.17 Find an example in which plurality with elimination violates IIA.

6.18 Which of the voting methods (plurality, Borda count, approval, range) violate the monotonicity condition? Explain.

6.19 This problem illustrates the IIA violation in the 1997 European Men's Figure Skating Championship. After skating, each skater receives a numerical rating from each of the seven judges that is converted to an *ordinal* ranking from that judge. Suppose that skater A's ordinals from the seven judges are 1, 1, 1, 3, 2, 2, 4; skater B's ordinals are 2, 3, 2, 1, 1, 4, 2; skater C's ordinals are 3, 2, 3, 2, 3, 1, 1. That is, the first judge ranked the skaters in the order A/B/C, the second judge ranked them in the order A/C/B, and so on. Find the winner using (a) plurality; (b) Borda count (with five candidates). The skating system was to find the median ordinal for each skater. (c) Show that each skater has median 2. The first tie-breaker was to count the number of voters who gave the median mark or better. (d) Show that skater C finishes third. The second tie-breaker is to add up the ordinals for the voters who gave the median mark or better. (e) Show that skater A wins. Now, suppose that there is one more skater to compete. This skater receives ordinals of 1, 4, 4, 4, 4, 2, 2. (f) Update the ordinals for skaters A, B, and C. (g) Show that the last skater finishes fourth, and determine the new top three (you will need a third tie-breaker, adding together all of the ordinals). (h) Explain why this is a violation of IIA, and discuss how the fans who watched the event would react to a shuffling of the order.

6.20 Give a positional voting representation for each voting system. (Assume that there are six candidates) (a) NFL MVP voting; (b) NBA MVP voting; (c) MLB Cy Young voting.

6.21 In an election with three candidates, 8 voters have the preference list A/B/C, 5 voters have B/C/A, and 4 voters have C/B/A. Under the positional voting system [x,2,1], find all values of x such that A wins under the positional voting system $[x, 2, 1]$. Find all values of x such that (a) A wins; (b) B wins; (c) C wins.

6.22 (a) Explain all four equations in Example 6.2. (b) Verify that the system of equations has an infinite number of solutions.

6.23 This exercise gives an alternative explanation of the PageRank equations. Imagine a bored person online who randomly clicks on links. (a) For the network in Example 6.2, site A can be linked to from sites B, C, and D. If the amount of time spent at the sites is named a, b, c, and d, respectively, explain why $a = \frac{1}{2}b + c + \frac{1}{3}d$. Interpret the solutions to Example 6.2 with this interpretation. (b) If a site has no links out, this interpretation falls apart. For the network of Example 6.2, suppose that site D links to site E, which has no links out. Write out the matrix for this network. (c) Now, imagine that the bored person types in a random url to end up at A, B, C, D, or E. Compare the resulting matrix to one where E links to all other pages. (d) Finally, suppose that 85% of the time the bored person clicks on a link and the other 15% of the time starts over with a preference for site A half the time, and B and C a quarter of the time. Write out the matrix-vector equation for this situation. This is close to the original PageRank system used by Google.

6.24 Using the linear equations from the text, find the probability of each upset in the NCAA basketball tournament. (a) #12 beats #5; (b) #9 beats #8; (c) #15 beats #2; (d) #8 beats #1; (e) #6 beats #3.

6.25 Ⓣ Pete Rose is not in the Baseball Hall of Fame due to his suspension for betting on baseball games. Barry Bonds is not in the Hall of Fame due to his involvement with steroids. Roger Maris had two MVP years but then had a string of injuries that cut his career short. Discuss your opinions on whether scandals of various types should disqualify someone from the Hall of Fame, and what the proper balance of historic performance and longevity should be.

6.26 Ⓣ Explain why Baseball Hall of Fame voting cannot violate the IIA criterion if each voter has an unlimited number of votes. However, suppose that a day before the election player A has enough votes for election and B does not. Then two other players are in the news, one in a positive way and the other in a negative way. Give a scenario in which these irrelevant players can cause B to be elected and A not.

6.27 Ⓣ In each of the following leagues, find a Condorcet cycle: NFL, NBA, MLB, college basketball, college football.

6.28 Ⓣ Suppose there are three basketball teams. Team A gives team B fits with its pressing defense, but team B gives team C problems with its outside shooting, and team C plays well against team A by breaking the press. Discuss how a Condorcet cycle could occur, and whether it makes sense to try to rank these teams.

6.29 Ⓣ For each method, propose a reasonable way to break ties. Hint: you may need to separately consider the cases where the number of voters are even or odd. (a) plurality; (b) Borda count.

6.30 Ⓣ If there is a Condorcet cycle, is it possible for there to be a Condorcet winner? Briefly explain.

6.31 Ⓣ Label each as True or False and briefly explain why. (a) If there is a Condorcet winner, then there is a majority winner. (b) If there is a majority winner, then there is a Condorcet winner. (c) If there is a Condorcet winner, that candidate will win the Borda count. (d) If there is a majority winner, that candidate will win the Borda count.

6.32 (T) If A and B are known to be the best of five candidates for an award, and you want A to win, how would you order your preference list to give A the best chance to win a Borda count? If approval voting is used, how would you vote to give A the best chance? If range voting is used (with a "star" system of 1, 2, 3, or 4) how would you vote to give A the best chance? Discuss the extent to which these voting systems are subject to strategic voting of this type.

6.33 (T) Read about slime mold in Ellenberg's *How Not to Be Wrong* and explain how slime mold violates IIA.

6.34 (T) In an episode of *Brain Games*, movie-goers showed a strong preference for a small tub of popcorn over a large tub. When a medium size was introduced (at a price close to the price of the large), customers chose the large tub over the small tub. Explain why this is a violation of IIA.

6.35 (T) Kenneth Arrow used the word "rational" to refer to the transitive property. Do you think that it is irrational to have a system that does not obey transitivity?

6.36 (T) Explain why the existence of Condorcet cycles proves that "majority rules" is not transitive.

6.37 (T) The concept of IIA is that a voting system's ranking of A and B should not be reversed based on the actions of some other candidate C. However, this means that the voting method should ignore the difference between ranking A first and B a close second compared to A first and B tenth. Do you think that IIA is a fairness condition that a voting system *must* obey?

6.38 (T) Suppose there are 121 voters in an election. (a) How many votes are needed to be a majority winner? (b) In the extreme, what is the fewest votes that a plurality winner could receive? (c) How many votes are needed to win with a 2/3 majority?

6.39 (T) Discuss the extent to which you think each of the following should be a factor in MVP voting. (a) playing on a winning team; (b) having the best statistics; (c) playing the full season; (d) being a good citizen.

6.40 (T) (a) In Figure 6.2, explain why the winning percentages decrease as vote uncertainty increases. (b) Explain why the winning percentages increase as the number of voters increases.

6.41 (T) (a) For range voting, give an example where one insincere or corrupt voter could change the outcome. (b) Explain why the trimmed mean option of dropping the highest and lowest scores is popular. (c) When evaluating the judges, is it a bad sign if a judge's score is often thrown out? never thrown out? Discuss how you might evaluate the judges.

6.42 (T) Discuss the difference between evaluating voting methods by whether they violate fairness conditions, versus evaluating voting methods by how "happy" the voters are with the decisions.

6.43 (T) In the commentary following Figure 6.7, explain why it is "cheating" to group together the 11- and 12-seeds.

6.44 (T) Discuss the advantages and disadvantages of being a 9-seed compared to an 11-seed in the NCAA basketball tournament.

6.45 Ⓣ In a round-robin tournament, each pair of teams plays exactly once. Suppose that two teams have the same number of wins. Show that there is a Condorcet cycle.

6.46 Ⓣ In a round-robin tournament, player A wins a medal if for every other player B, either A beat B or A beat a player who beat B. (a) Show that at least one player wins a medal, and (b) find a circumstance in which very player wins a medal.

6.47 Ⓒ In the 1989-90 NBA MVP voting, 92 voters awarded 27 first-place votes to Magic Johnson, 38 to Charles Barkley, and 21 to Michael Jordan. Johnson won with 636 points. Suppose that Barkley only received first-place and third-place votes. How many voters must have left Barkley off the ballot completely?

6.48 Ⓒ Suppose that the voters have a single issue on which they will base their votes. In this case, some possible preference lists cannot occur (one of Arrow's assumptions is that all lists are possible). Think of a candidate's position on the issue as being marked on a football field, ranging from left end zone to right end zone. Each voter's stance is also represented by a yard-line, and the preference list is determined by how close each candidate is to that yard-line. If candidate A is at the left 10 yard-line, B is at the left 40 yard-line, C is at the right 30 yard-line, and you are at the 50 yard-line, your preference list will be B/C/A since B is the closest to you, followed by C. Explain why the preference lists A/C/B and C/A/B cannot occur. In this situation, show that violations of IIA cannot occur.

6.49 Ⓒ Suppose that there are only two candidates and an odd number of voters. Explain why there will be a majority winner, and show that the majority winner will also win plurality and the Borda count. That is, majority rules works when there are two candidates. Show that approval voting and range voting will not necessarily elect the majority winner, however.

6.50 Ⓒ Six dice are numbered in unusual ways. Die A has all 3s; die B has four 4s and two 0s; die C has three 5s and three 1s; die D has two 6s and four 2s. Show that the dice violate transitivity, in that A loses to B, which loses to C, which loses to D, which loses to A.

6.51 Ⓒ For range voting on a scale of 1 to 5, if a player currently has a range average of 4.6 and you think the average should be 4.0, can rating the player 1 accomplish your goal? Determine the number of votes for which this could be true.

6.52 Ⓒ A small network has 5 sites. A links to B and D; B links to A, C, and E; C links to B and D; D links to A, B, and E; E links to A and C. Write out the basic rating matrix (as in Example 6.2) for this network and solve it.

6.53 Ⓒ A team has five players. During the course of a game, A passes to B 14 times, to C 22 times, to D 18 times, and to E 6 times. B passes to A 10 times, to C 8 times, to D 12 times, and to E 14 times. C passes to A 4 times, to B 8 times, to D 3 times, and to E 5 times. D passes to A 16 times, to B 6 times, to C 22 times, and to E 14 times. E passes to A 13 times, to B 12 times, to C 11 times, and to D 10 times. Set up the PageRank matrix for the passer ratings and reduce the matrix.

6.54 Ⓒ In *Analyzing Wimbledon*, Klaasen and Magnus analyze the success of different seeds not by number (as we did in this chapter) but by expected round reached. With 128 players, there are seven rounds. The 1-seed should win 7 matches, the 2-seed should win 6, both the 3- and 4-seeds should win 5, seeds 5-8 should win 4, and so on. Discuss how this compares to computing the log base 2 of the seedings and rounding up.

6.55 (P) Develop a program which will take two teams A and B and find the shortest link of teams such that A beat C beat D beat B. What percentage of teams can be linked this way? Identify two teams to be in the same group if there is a link from A to B and a link from B to A. How many different groups are there?

6.56 (P) Run the simulation discussed in the section. For each of 1000 voters, generate the voter's rating of player 1 (50 plus a sample from a normal distribution with mean 0 and standard deviation 1) and player 2 (49 plus noise) and so on, compute the rank and hence the points assigned to that player by that voter, and add up the points for each player to determine the voting method's ranking of the players. For each pair of players, determine which player was preferred by a majority of voters; if the voting method ranks that player higher, give the voting method a "win." If there is no majority winner, do not count the matchup as a win or a loss. Do this for a variety of voting methods, initial values of the players, and standard deviations.

6.57 (P) Run the PageRank system on the sports league of your choice. Compare the rankings to those obtained from other systems and critique the PageRank system.

Further Reading

Amy Langville and Carl Meyer's *Who's #1?* is an excellent resource for the ratings systems covered here plus several more.

The Heisman vote totals are from the Heisman web site: http://heisman.com/roster.aspx?rp_id=74&path=football accessed 4/9/15.

The Barry Wilner quote is from http://profootballtalk.nbcsports.com/2011/12/28/ap-explains-voting-process-for-nfl-awards/ accessed 4/9/15.

NFL MVP voting history is from http://www.coldhardfootballfacts.com/content/voting-history-why-peyton-manning-should-run-away-with-nfl-mvp -award/20626/ accessed 4/9/15.

The NBA MVP voting history can be found at http://www.basketball-reference.com/awards/ accessed 4/9/15.

The MLB MVP and Cy Young voting history can be found at http://www.baseball-reference.com/awards/ accessed 4/9/15.

There are several good books about voting methods. Kenneth Arrow's *Social Choice and Individual Values* is an early summary of Arrow's work. Jeff Suzuki's *Constitutional Calculus* is excellent. Stories about Lewis Carroll and voting are in William Poundstone's *Gaming the Vote*. A journal article of note is in the October 2009 *American Mathematical Monthly* by Daugherty, Eustus, Minton, and Orrison.

Google's description of PageRank with helpful links can be found at http://google.about.com/od/searchengineoptimization/a/pagerankexplain.htm accessed 5/27/15.

The June 2014 issue of the *Journal of Quantitative Analysis in Sports* has an article that adapts the PageRank algorithm to an effective rating system. See "An Oracle method to predict NFL games" by Balreira, Miuli, and Tegtmeyer.

The first issue in 2015 (volume 11, number 1) of the *Journal of Quantitative Analysis in Sports* has several excellent articles about March Madness. An analysis of the 5-12 matchup can be found at FiveThirtyEight: http://fivethirtyeight.com/tag/no-12-seeds/ accessed 4/9/15. Peter Keating's *Giant Killers* talk at the 2014 Sloan Sports Analytics Conference is interesting.

Chapter 7

Saber- and Other Metrics

Introduction

FIGURE 7.1: Bill James

In the movie *Moneyball* the character Peter Brand, played by Jonah Hill, makes the case for the use of statistical analysis in baseball, "Using stats the way we read them, we'll find value in players that nobody else can see. People are overlooked for a variety of biased reasons and perceived flaws: age, appearance, personality. Bill James and mathematics cut straight through that."

The one-sentence sound-bite review of *Moneyball* is that it describes the conflict between academic stat-head analysts and old-school baseball insiders. Such a review is overly simplistic, but the basic conflict is real, and is a recurring theme of modern society. As Ian Ayres puts it in his book *SuperCrunchers*, "We are in a historic moment of horse-versus-locomotive competition, where intuitive and experiential expertise is losing out time and again to number crunching." However, unless you're selling books or movies you should not believe that this is an us-versus-them battle.

If you carefully read Peter Brand's speech, you can identify the strengths and weaknesses of the use of statistics in sports. A baseball player's playing statistics do not care if the player looks overweight or has an odd technique. The analyst can "cut straight through" those biases. However, the numbers may also overlook the odd technique that is likely to lead to injury, or the temper that makes the player uncoachable. Smart users of analytics (including the Oakland A's of Billy Beane, who now have a larger scouting budget than they did before *Moneyball*) draw on all information to make decisions. Analytics should not be ignored *or* blindly followed.

The word "analytics" will appear frequently in the remaining four chapters. For some, analytics only refers to the products of sophisticated statistical analysis. In this book, **analytics** will refer to the identification of patterns in data through any statistical analysis.

In this chapter, we take a first look at **sabermetrics** (baseball analytics) and analytics in other sports. We will answer the following questions. How do points scored and allowed relate to wins? What makes a statistic useful? What are some of the useful statistics that have been developed recently? How can defense be measured? How can the effect of a distinctive home field be measured? Is it possible to apply the new statistics to players in the past?

The Pythagorean Cult

Pythagoras (c.571-496 B.C.) was a Greek mathematician and philosopher. He and his followers are credited with numerous mathematical discoveries and advances in music, astronomy, and other fields of inquiry. The Pythagoreans were a secretive bunch, so we know distressingly little of what they actually did. The theorem that bears Pythagoras' name (the relationship $a^2 + b^2 = c^2$ holds in any right triangle of side lengths a, b, and c) was not original to the Pythagoreans. Its connection to baseball remained hidden for centuries.

Bill James (1949-) is for many people the father of modern sports analytics. He did his original work as a night watchman at a pork-and-beans factory. There are two relevant facts here: he was a baseball outsider, and he was dedicated enough to take a job that gave him time to copy and analyze a season's worth of box scores by hand. He self-published results in what became a highly influential series of books, the *Bill James Baseball Abstract* series (1977-1988). Because of his outsider status, his ideas and their evolution were very publicly aired and allowed a generation of sports fans to develop ideas along with him. Among his first ideas was what he called the **Pythagorean Method**.

The philosophy behind the Pythagorean Method is crucial. And simple. The most important statistic for a baseball team is wins. It follows that everything else we do statistically should relate to the ability to win games. A statistic should always relate to wins in a measurable way. Backing up one step, the way you win games is by scoring more runs than the opposition. If we can relate runs scored to wins, then any statistic that can be related to runs can be related to wins. The first step is to find an equation that relates runs and wins. What Bill James found is this: if a team scores RS runs and allows RA runs, then its winning percentage (more precisely, winning *proportion*) will be approximately

$$\mathrm{WP} = \frac{\mathrm{RS}^2}{\mathrm{RS}^2 + \mathrm{RA}^2}.$$

I assume that James chose the name Pythagorean Method because of the sums of squares in the denominator.

Example 7.1

	W	L	RS	RA
SF	88	74	665	614
KC	89	73	651	624

For the participants in the 2014 World Series, compute the expected wins using the Pythagorean Method and compare to the actual wins.

Solution. The table indicates that San Francisco scored 665 runs and allowed 614. The Pythagorean Method predicts a winning proportion of $665^2/(665^2 + 614^2) = 0.5398$. Multiplied by 162 games, the Giants are predicted to win 87.45 games, very close to the actual total of 88 wins. For Kansas City, the Pythagorean Method predicts a winning proportion of $651^2/(651^2 + 624^2) = 0.5212$. Multiplied by 162 games, the Royals are predicted to win 84.43 games, less than the actual 89 wins.

The exponent 2 is used in Example 7.1, but other exponents a in the general formula $\text{WP} = \dfrac{\text{RS}^a}{\text{RS}^a + \text{RA}^a}$ are equally valid theoretically. Which exponent is best? In the absence of theory, the way to answer the question is to try several exponents and see which one makes the best predictions. By "best predictions" we mean the smallest differences between actual wins and predicted wins. A common way of computing a total error is to add together the squares of the individual team errors.

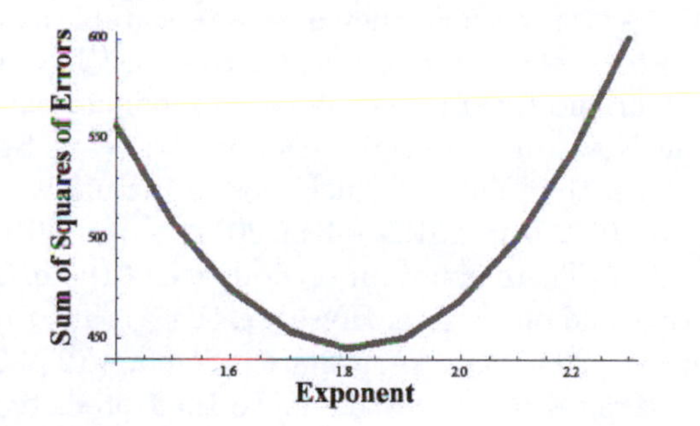

FIGURE 7.2: The Best Exponent, MLB 2014

Example 7.2 Make a table of errors in Pythagorean Method predictions for the 30 teams in 2014. Compute the sum of the squares of the errors. Then repeat this for exponents $a = 1.4$, $a = 1.5$, ..., $a = 2.3$ and find the exponent that minimizes the sum of the squares of the errors.

Solution. If the error is computed as actual wins minus predicted wins, the errors in Example 7.1 are $88 - 87.45 = 0.55$ and $89 - 84.43 = 4.57$. The sum of the squares of all of the errors with the exponent 2 equals 468.9. This can be computed in Excel, for example. Changing the exponent in Excel is easy and quick. The first few lines of a table of errors is shown below. Data are from baseball-reference.com. All teams played 162 games.

	W	RS	RA	PY	error	square
ARI	64	615	742	65.97	-1.97	3.88
ATL	79	573	597	77.68	1.32	1.75
BAL	96	705	593	94.88	1.12	1.26
BOS	71	634	715	71.31	-0.31	0.09
CHC	73	614	707	69.65	3.35	11.21
...	..	...	...	...	...	...

The total error is the sum of all of the entries in the last column shown. Figure 7.2 shows the sums of squares of errors for the other exponents. To one decimal, the minimum is 445.3 corresponding to the exponent 1.8.

Theoretically, it can be shown that under certain assumptions the winning proportion is $\text{WP} = \dfrac{\text{RS}^a}{\text{RS}^a + \text{RA}^a}$ for some exponent a. The exponent $a = 1.82$ is often cited as the best for baseball; for 2014, $a = 1.81$ is actually slightly better. Although the scale in Figure 7.2 has been magnified, notice that the exponent $a = 2$ works reasonably well for baseball. Exponents for other sports are found in the exercises.

Pythagorean Method predictions are listed at baseball-reference.com and other sites. The error is commonly labeled as "luck." The label implies that there is no rhyme or reason to the errors. Kansas City won 4 more games than predicted in 2014; is this because they have a repeatable skill for winning close games (such as great relief pitching) or did they just have good luck?

Given that Mariano Rivera was the most dominant closer in baseball history, perhaps the New York Yankees are a good test case. Starting in 2013 and working backwards, the Yankees' "luck" value (actual wins minus predicted wins) was +6 in 2013, 0 in 2012, −4 in 2011, −2 in 2010, +8 in 2009, +2 in 2008, −3 in 2007. There is not an obvious trend there. Three are positive, three are negative, and one is zero. However, there is a reason I stopped where I did. The values for 2006 back to 1998 are +2, +5, +12, +5, +4, +6, +2, +2, and +6. They averaged nearly 5 wins more than predicted for 9 years! This does not look like luck.

Most teams follow a random-looking up/down pattern of errors, so the phrase "luck" is appropriate. As with most analytics, however, be aware that there are exceptions to the rule. The physics of sports deals with events that *must always* occur. Analytics uncovers interesting patterns in events that have happened. The patterns may or may not continue to hold in the future.

Stanley Rothman has provided an interesting alternative to the Pythagorean Method. He asked a simple question; why does it have to be so complicated? Many patterns are linear or approximately linear, perhaps winning percentage is one of them. He proposes the equation $\text{WP} = 0.000683(\text{RS} - \text{RA}) + 0.5$. The sum of the square errors for 2014 is 407.4, less than the Pythagorean sum of square errors of 445.3. The average absolute error of the linear model is also smaller, at 2.82 compared to 2.89. Other than the constant 0.000683 being ugly-looking, this model is an improvement on the Pythagorean Method!

When Good Statistics Go Bad

We have established a strong connection between runs in baseball and wins. Similar connections can be made between points and wins in sports like basketball and football, or goals and wins in soccer and hockey. The next step is to relate individual statistics to runs (or points or goals). Since most of the early work was done in baseball, we develop this idea with baseball statistics.

Some basic definitions may help. **Batting average** is computed as hits divided by at bats. Three decimal places are traditional, and an average of .406 is pronounced "four oh six" and is an average that is "over four hundred." The number of at bats does not count plate appearances in which the batter walks, is hit by a pitch, or is credited with a sacrifice. **On-base percentage** (OBP), by contrast, equals hits plus walks plus hit by pitches divided by total plate appearances. For both statistics, a batter who reaches base due to a defensive error is debited with an out.

Batting average is a statistic that dates back to the beginnings of the game. Generations of fans have learned that a batting average of .300 is excellent and batters near the "Mendoza line" of .200 are not long for the major leagues. Nevertheless, batting average as a measure of batting performance was one of the first ideas that Bill James challenged. If batting average does not have a strong connection to runs scored, then its worth as a statistic is dubious. A **scatter plot** is the graph of a set of points representing two variables. In the scatter plot to the left in Figure 7.3, a point (x, y) consists of a team's batting average in 2004 as x and the number of runs the team scored in 2004 as y.

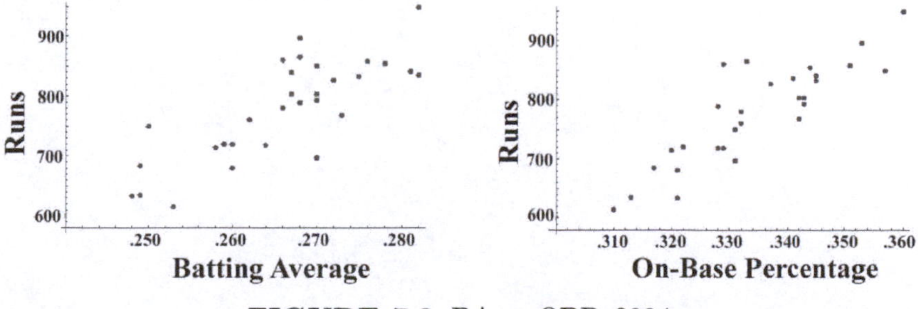

FIGURE 7.3: BA vs OBP, 2004

In both plots, there is a clear trend that points to the right are higher. That is, teams with better batting averages (or better on-base percentage) scored more runs. This is not surprising, and indicates that both statistics are reasonable measures of run-scoring ability, and hence winning ability. Which statistic is better? We want the best relationship possible to runs scored. Try an example: predict the runs scored for a team with batting average .270. In the scatter plot, the points above the .270 label show run values from 700 to 850; this is not a very specific prediction. We would like a smaller range of values, with the points more tightly grouped vertically. The closer the points come to a straight line, the better.

Visually, it may or may not be clear that on-base percentage gives tighter predictions. There are ways to quantify this (and avoid any optical illusions from the graph). Both regression and correlation will be discussed in later sections. Linear regression is easy enough to visualize in Figure 7.3 as a line through the center of the data. This line can be thought of as predicting values of runs scored, so we can use the sum of the squares of errors (SSE) to evaluate the predictions. Predictions from batting average have SSE equal to 71539, while predictions from on-base percentage have SSE equal to 47159. We get much better predictions of runs scored (and hence wins) from on-base percentage! Correlation indicates how strong a linear relation is; in this case, the larger the correlation, the stronger the relationship. The correlation between batting average and runs scored is 0.80, while the correlation between on-base percentage and runs scored is 0.88.

For these reasons, on-base percentage became known as a better measure of batting performance than batting average. On-base percentage was a hallmark of the original *Moneyball* movement, as players who could draw walks became more valued and batters were taught to take more pitches (and, in the process, tire out the opposing pitcher). An even better batting metric is OPS, short for On-base Plus Slugging (slugging percentage equals total bases divided by at bats; total bases, in turn, counts 4 for a home run, 3 for a triple, and so on). It is considered better because it has a stronger relationship to runs scored. The scatter plot in Figure 7.4 (from 2014 data) is clearly tighter than either of those in Figure 7.3.

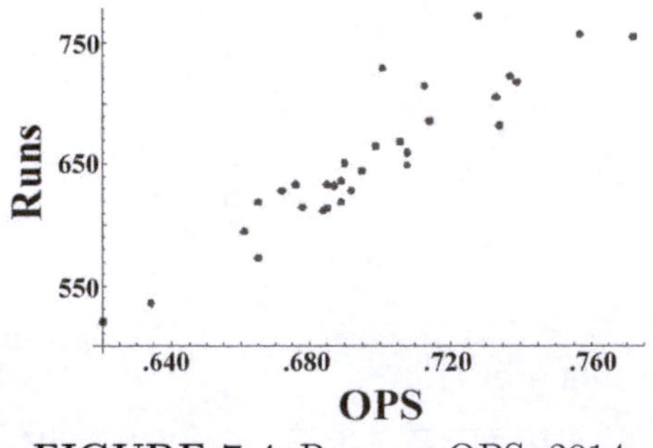

FIGURE 7.4: Runs vs OPS, 2014

The new and better statistics are empirical, meaning that they are data-driven. An important implication of this is that evaluations may change when the data change. Compare Figure 7.3 with its 2014 analog in Figure 7.5. Check the vertical scale of runs scored to see how much scoring decreased in a decade. In these plots, it is hard to tell any difference between batting average and on-base percentage. In fact, the SSE values are nearly identical (33518 for batting average versus 33458 for on-base percentage) and the correlations are nearly identical (both at 0.797 to three decimals). Compare these to the OPS statistic, which has SSE equal to 16487 and a correlation to runs scored of 0.91.

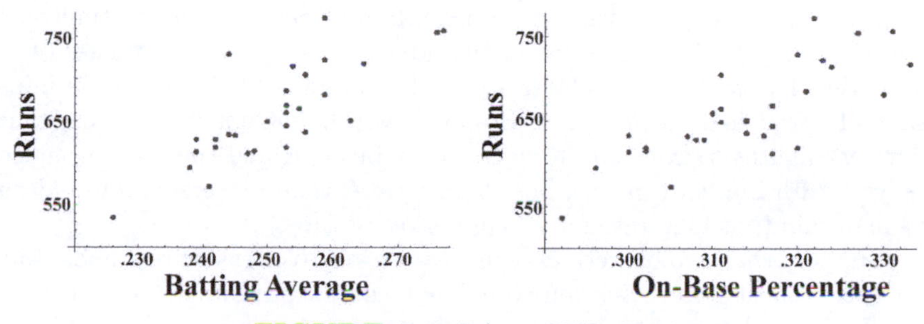

FIGURE 7.5: BA vs OBP, 2014

In 2014, then, OBP holds no clear predictive advantage over batting average. The new truths of one generation may become the outdated dogma of another generation.

Rates versus Numbers

Many of the best-known baseball statistics are **counting statistics**. These include home runs, RBI, and strikeouts, where you simply count how many times each thing happened. By contrast, batting average and on-base percentage are **rate statistics** computed as ratios. Batting average, for example, is measured in hits *per at bat*. In general, counting statistics measure total production and are highly influenced by how often a player is in position to achieve the statistic. For example, a basketball player could be a great shooter but will not score much if most of the shots are taken by teammates. Rate statistics are more about efficiency, typically measuring how often an event happens compared to the number of opportunities.

The most popular basketball statistic is points scored, which is a counting statistic. While points per game is a rate statistic, the number of games played is not a good representation of the player's opportunities. Points per minute is fairer to the player who gets limited minutes per game. Computing points per 48 minutes puts the statistic on a familiar scale where 30 is a large value.

Example 7.3 Player A scored 28 points in 40 minutes and player B scored 12 points in 14 minutes. Compute points per 48 minutes for each player.
Solution. Player A scored $\frac{28 \text{ pts}}{40 \text{ min}} = 0.7$ points per minute. Multiply by 48 minutes to get 33.6 points per 48 minutes. Player B scored $\frac{12}{14}(48) = 41.1$ points per 48 minutes. If player B had maintained the same scoring rate for 48 minutes, he or she would have scored 41 points.

Player B scored points at a faster rate than player A, given the time on the court. Example 7.3 shows some of the advantages and disadvantages of rate statistics. Knowing that a player scored 12 points might disguise the impact that player B had on the game. However, you might not want to use points per 48 minutes to conclude that B had a better game than A. In general, players with limited time can post higher rates than players who would have to maintain that high rate for a long period of time.

Suppose that two basketball teams each grab 10 offensive rebounds. Based on this counting statistic, you would not see an advantage for either side. Example 7.4 shows where a rate statistic uncovers a strength that could have been missed.

Example 7.4 Team A made 45 out of 80 shots, while team B made 35 out of 80 shots. Both teams have 10 offensive rebounds. Compute the offensive rebounding rate for each team.
Solution. Team A missed 35 shots, and got offensive rebounds on $\frac{10}{35} = 0.27$ rebounds per opportunity. Team B missed 45 shots, and got offensive rebounds on $\frac{10}{45} = 0.22$ rebounds per opportunity. Team A did a better job of rebounding its own misses.

Persistence and Reliability

Many statistics are designed to measure a particular skill. In basketball, shooting is measured by shooting percentage, rebounding by total rebounds or rebounding percentage (as in Example 7.4), passing by assists, and so on. Is there any way to tell whether a statistic actually captures a player's ability? One way of analyzing what a statistic measures is to check for consistency, what is variously called **persistence** or **reliability**.

Suppose that we invent a new statistic that is designed to measure clutch play, the ability to perform well at the most important times. We compile our list of top 10 clutch players each year, but nobody ever makes the list more than once. Year after year, players' clutch ratings change in a seemingly random pattern. There are two possibilities. One is that there is no such thing as clutch play; there may not be a skill there to measure. The other possibility is that our statistic does not measure what we want. Our statistic may be influenced by other skills at cross-purposes to clutch play, or our statistic might be allowing in too much noise. Either way, we need to modify or abandon our statistic.

A good skill statistic will be stable, in that a player or team who rates highly one year should usually rate highly the next year. There are injuries and other influences that will cause some change, but generally the best ratings should persist. A way of testing this is to compile the statistics for consecutive years for all players or teams who played both years. A high correlation between consecutive years is a good sign. **Correlation** is a measure of the linear relationship between two variables. If there is an exact linear relationship (for example, next year's value is exactly the same as this year's, or exactly half), then the correlation equals 1, and we have high confidence that we can use this year's value to predict what next year's value will be. If the two variables are not related at all, the correlation will equal 0, and we have zero predictive ability. (The converse is not true, so a correlation near 0 does not prove that the variables are unrelated.) If the variables are related through a line with negative slope (so large values one year predict small values the next year), the correlation is -1. The correlation is always between -1 and 1.

A technical definition of correlation may aid those familiar with vector analysis. If the n values of variable x are collected in a vector $\mathbf{a}$ and the n values of variable y are collected in a vector $\mathbf{b}$, then the correlation between x and y equals the dot product of $\mathbf{a}$ and $\mathbf{b}$ divided by the magnitudes of $\mathbf{a}$ and $\mathbf{b}$. Stated differently, the correlation is the cosine of the angle between the vectors: if the vectors line up in the same direction, the correlation is 1. The opposite direction gives a correlation of -1. A correlation of 0 follows from a 90-degree angle, so that the vectors are perpendicular (or orthogonal, as we like to say).

Baseball Between the Numbers, from the *Baseball Prospectus* group, compiled such year-to-year correlations for a variety of pitching statistics. Strikeouts per batter faced (note the use of a rate statistic) had a year-to-year correlation (using data from 1972 to 2004) of 0.790. Walks per batter was less consistent at 0.676, then hits allowed per batter at 0.499, home runs per batter at 0.470, ERA (earned run average) at 0.380, and winning percentage at 0.204. The takeaway is not that winning percentage for pitchers is useless, but that it clearly depends on many more factors than just the pitcher's performance. In fact, attempts to find pitching "winners" who win close games by grit and iron will have failed as thoroughly as attempts to find clutch hitters. Strikeouts, on the other hand, are very much under the control of the pitcher and the high correlation shows that high (or low) strikeout rates tend to be maintained by pitchers year after year. It makes sense to call a pitcher a "strikeout pitcher" or a "groundball pitcher" (percentage of ground balls has a year-to-year correlation of 0.807), but calling a pitcher a winner is not very meaningful.

On the Defensive

There is an old sports adage that "defense wins championships." We will see in the exercises that this is not especially true. It may be that the adage is mostly a nod to players who play good defense in sports where defensive ability is hard to measure. Baseball is the major sport in which defense is close to an individual endeavor, so we should be able to evaluate defensive ability in baseball. And yet, we have the case of Derek Jeter, who won five Gold Glove Awards as the best shortstop in the American League, but is consistently rated as below average by sophisticated defensive ratings.

Jeter typically had a good fielding percentage. This is the original fielding statistic, and is defined as the ratio of successful plays (assists plus putouts) to chances (assists plus putouts plus errors). Errors are somewhat of a judgment call, but require the fielder to get his glove on the ball. A slow fielder who can't even reach a ground or fly ball is not penalized, while a speedy fielder who tracks down a ball that nobody else would get to is penalized if he then makes an error on, for example, the throw. The effect can be seen in Example 7.5.

Example 7.5 Player A is a shortstop with 273 putouts, 392 assists, and 13 errors. Compute his fielding percentage. Player B is a little faster and reaches 28 more balls, leading to 27 more assists and 1 more error. Compute his fielding percentage. Determine how many assists are needed to compensate for one error.
Solution. Player A has a fielding percentage of $\frac{273+392}{273+392+13} = \frac{665}{678} = 0.9808$. Adding 27 more plays to the numerator and 28 more chances to the denominator, his fielding percentage is now $\frac{692}{706} = 0.9802$, which is less than before. Adding another 24 successful plays gives a fielding percentage of $\frac{716}{730} = 0.9808$. That is, player A would need to make 51 successful plays to offset one error in the fielding percentage statistic.

The question to ask is whether you would rather have player B or A. Most would take B, even though his fielding percentage is worse. Player B made 27 more successful plays, a full game worth of outs. Sometimes runners advance more on errors than on base hits, but often the impact of an error is identical to that of a hit. Would you trade 27 outs for one hit? The starting numbers are, perhaps, extreme but are taken from Derek Jeter's 2004 season.

An early attempt at improving on fielding percentage is Bill James's **Range Factor**, which equals putouts plus assists divided by games played. The simple idea is that the more plays you make, the better you are. An error is no different than being too slow to get to the ball in the first place. To take one example, in 2004 Derek Jeter won a Gold Glove at shortstop. His fielding percentage of 0.9808 was fourth in the league among shortstops who played more than 100 games, and his 273 putouts led the league. Yet, his Range

Factor of 4.46 plays per 9 innings was below the league average of 4.56 plays per 9 innings.

With new tracking data, discussed in Chapter 10, major improvements are being made in defensive statistics.

Plus and Minus

In many sports, defense is very hard to measure because individual responsibilities are impossible to define. A basketball player gets a layup off of a pick-and-roll. Does the fault lie with the player "guarding" this player? Perhaps his or her role was to stop the dribbler and the pass was open because the other defender did not switch to the roll player fast enough. But, where was the help defense? And which side was the help to come from? It may be impossible to assign blame correctly.

These types of plays in basketball and most of the action in hockey, football, and soccer are so fluid that they are hard to evaluate. An idea that has gained some traction is the **plus-minus** statistic, which attempts to divide credit or blame for points. The basic form of the plus-minus for a player is the net score (team points for minus points against) while that player is in action (on the court, on the ice, and so on). If a five-person group outscores the opposition by two, then each of the five people scores +2, while each of the opponents (assuming no substitutions) scores −2.

The obvious flaw to the plus-minus is that a player on a bad team will almost surely have a negative rating, no matter how good the player is. Brian Burke famously expressed this in more colorful language at the 2014 Sloan Sports Analytics Conference. (The session had the impish title "Hockey Analytics: Out of the Ice Age.") A partial fix is to compute the difference of the team's net score when the player is in the game and out of the game. A player on a bad team gets a +5 rating if the team is "only" outscored by 8 points when the player is in the game compared to being outscored by 13 when the player is out of the game. As a method for evaluating individual players, this can still be unsatisfying. If a mediocre player is always put in the game by the coach when the team superstar is playing, the mediocre player will look good.

If the lineups vary enough, there are ratings methods that can assign individual contributions of players to each lineup. For example, in basketball each five-person group can be treated as a distinct team, and a rating method such as those discussed in Chapter 5 can be used to rate each team. An individual player's rating could be the sum of the group ratings for all of the groups to which that player belongs, weighted by the number of minutes that the groups played together and compared to the overall team rating.

Plus-minus ratings are used by coaches to evaluate the effectiveness of different groupings of players.

Park Factors

A basketball court is 94 feet long, a hockey rink is 200 feet long, a tennis court is 78 feet long, and so on. By contrast, baseball and golf playing fields have dramatically different sizes and shapes. Further, outdoor sports can be strongly influenced by environmental issues such as temperature and altitude. For this reason, baseball in particular has a need for corrections for the site of events. The statistics for someone playing half of his games in Coors Field can and do look different from someone whose home field is Safeco Park in Seattle.

The proper way to correct for home field in baseball is not entirely simple. Suppose that 50% more home runs are hit in Fenway Park than in any other stadium. Before you conclude that Fenway Park is 50% easier to hit home runs in than average, think about which teams play there. If the Red Sox are loaded with home run hitters, then the extra home runs in Fenway could be a result of Red Sox ability instead of the park. Also, the result could be due to the generosity of the Red Sox pitchers. We need to factor in Red Sox games both in Fenway and away from Fenway to get a read on how much the park affects home runs. The following gives a simple method of computing park factors.

If H_h is the number of home runs that team A hit at home, H_a is the number of home runs that team A allowed at home, R_h is the number of home runs that team A hit on the road, and R_a is the number of home runs that team A allowed on the road, then team A's park factor for home runs is given by

$$\mathrm{PF} = \frac{(\mathrm{H}_h + \mathrm{H}_a)/\mathrm{H}_g}{(\mathrm{R}_h + \mathrm{R}_a)/\mathrm{R}_g}$$

where H_g is the number of home games that team A played and R_g is the number of road games played. In a full season, $\mathrm{H}_g = \mathrm{R}_g = 81$, and the park factor simplifies to the number of home runs hit (by either team) in team A's home games divided by the number of home runs hit in team A's road games. In 2014, Yankee Stadium had the largest park factor for home runs at 1.47 and AT&T Park in San Francisco had the lowest at 0.68. (Data from espn.com.)

You may be surprised that Coors Field did not have the largest home run park factor (it was second at 1.39). The reason indicates why park factors for other statistics need to be computed. With outfield fence dimensions of 347 feet to left, 410 feet to center, and 350 feet to right, Coors Field has one of the largest fields in baseball. The fences need to be pushed back to reduce the number of home runs. However, the spacious outfield allows more balls to fall for hits: the 2014 Coors Field park factor for hits is 1.32 (second place is 1.08). Another factor is the amount of foul ground: parks with the seats on top of the field do not allow fielders to catch many pop fouls for outs, giving hitters extra pitches to face. When you put it all together, Coors Field has the largest runs park factor at 1.50 (50% more runs are scored at Coors Field!), with Chase Field in Phoenix second at 1.15 and Safeco Park lowest at 0.82.

With this method of computing park factors, a player's statistics can be park-adjusted in a number of ways. Since Coors Field multiplies home run totals by 1.39, you might think that a Colorado player should have his home run total divided by 1.39. This, however, assumes that *all* of the player's games are at Coors. Instead, suppose a player played half of his games in Coors Field and the other half in parks for which the average park factor is 1. Then *half* of his home runs should be divided by 1.39. In general, for a player whose home field has park factor pf, the park-corrected value should be $\frac{2x}{1+pf}$ where x is the uncorrected statistic. In the case of Coors Field home runs, notice that $\frac{2}{1+pf} = \frac{1}{1.195}$. Instead of dividing out the 39% increase that Coors Field provides, we divide out the 19.5% increase that playing half of his games in Coors Field provides.

Example 7.6 In 2014, Nelson Cruz led the American League with 40 home runs for Baltimore. Chris Carter of Houston was second with 37. Given home run park factors of 0.936 for Baltimore and 1.173 for Houston, compute the park-adjusted home run numbers. In 2014, Clayton Kershaw led the National League in ERA at 1.77 for Los Angeles. Johnny Cueto of Cincinnati was second at 2.25. Given run park factors of 0.907 for Los Angeles and 0.963 for Cincinnati, compute the park-adjusted ERA numbers.

Solution. We can see that Cruz played in a tough park for home runs, while Carter had it relatively easy. The park-adjusted value for Cruz is $\frac{80}{1.936} = 41$, while the park-adjusted value for Carter is $\frac{74}{2.173} = 34$. Cruz earned his home run championship. For the pitchers, both Kershaw and Cueto played in pitcher-friendly parks. The park-adjusted value for Kershaw is $\frac{3.54}{1.907} = 1.86$ and the park-adjusted value for Cueto is $\frac{4.5}{1.963} = 2.29$. Kershaw still has a sizable lead, but the lead is not as large.

You should object to our use of the same adjustment model for pitchers and batters. The assumption behind the model is that half the games are played at home and half on the road at parks that have an average park factor of 1. For batters who play every day, this is not a bad assumption, unless their team plays in a division with unusually high or low park factors. Pitchers, who only pitch in every fifth game or so, could easily have an unequal distribution of parks pitched in. It would be more accurate in all cases to weight the adjustments by how many games are played in each park.

You might consider a further tweak to the park factor model, trying to account for different parks being better or worse for different players. Fenway Park is great for lefthanded line drive hitters who have would-be outs bounce off the Green Monster for doubles, while righthanded line drive hitters whose would-be home runs bounce off the Monster are hurt by Fenway Park. (The doubles park factor for Fenway Park in 2014 was 1.523, by far the largest in the major leagues.)

Four Factors, Fenwick, and Football

In the exercises, you will be asked to explore questions about a variety of sports, with suggested websites to find relevant information. Since the text to this point has been baseball-heavy, in this section we look at a couple of ideas from other sports.

Dean Oliver's **Four Factors** evaluate basketball team performances. For our purposes, they also illustrate an important guide to developing ratings. Among statistics that are readily available, which ones identify the better team? Points scored is the obvious answer, but it's not a helpful answer. Another unhelpful, though correct, answer is "all of them." To give an example of why this is unhelpful, suppose you have numbers for offensive rebounds, defensive rebounds, and total rebounds. You don't want to use all three in your analysis: if you have two of them, it is not hard to figure out the third. The more interesting question is whether offensive rebounds are more or less important than defensive rebounds. And, by the way, we should remember to frame the statistics as rebound rates and not just raw numbers.

In order of importance, Oliver's four factors are shooting, turnovers, rebounding, and free throws. There is nothing surprising here, but the trick is to define each category in a way that extracts useful information without duplicating information from other categories. As discussed in Example 7.4, shooting and rebounding can overlap. Here is one version of the Four Factors.

Shooting is measured by **effective field goal percentage** where 3-point shots made are given 50% more credit (since they are worth 50% more points). Then we compute $S = \frac{\text{FGM}+0.5\text{TPF}}{\text{FGA}}$ where the team attempted FGA field goals and made FGM field goals, of which TPF were three-pointers. A team's shooting rating for a game is the difference in S-values for that team and its opponent.

Turnovers is a rate statistic, with the denominator being the number of possessions. Since the number of possessions is not a statistic that is normally available, we use the estimate $\text{FGA} - \text{OREB} + 0.4\text{FTA} + \text{TO}$ where OREB is offensive rebounds and free throws attempted (FTA) is multiplied by 0.4 to compensate for multiple free throws per possession. Typically, 2 free throws constitute a possession, but "and-one" free throws do not represent a new possession and sometimes a player shoots three free throws. On average, 0.4 (some use 0.44) seems to work well. The team's turnover rating is the difference in its rating and its opponent's rating of $T = \frac{\text{TO}}{\text{FGA}-\text{OREB}+0.4\text{FTA}+\text{TO}}$.

Rebounding is a rate statistic, with offensive rebounding measured by $\frac{\text{OREB}}{\text{OREB}+\text{DREB}opp}$ where DREB*opp* is the number of defensive rebounds by the opponents. The denominator could be slightly different from the number of missed shots FGA−FGM. Defensive rebounding is the complement of the opponents' offensive rebounding rating, so computing the difference in offensive rebound ratings accounts for all rebounds.

Finally, free throws are measured as a rate statistic. Instead of looking at free throws per possession, Oliver divides free throws made by the number of field goal attempts to compute $\frac{\text{FTM}}{\text{FGA}}$.

Example 7.7 In a 2015 playoff game, San Antonio made 42 of 91 field goal attempts, 8 of them three-pointers, made 19 free throws, had 10 offensive and 38 defensive rebounds, and had 9 turnovers. Los Angeles made 39 of 92 field goal attempts, 9 of them three-pointers, made 20 free throws, had 16 offensive and 39 defensive rebounds, and had 11 turnovers. Compute the Four Factors for each team.

Solution (1) San Antonio's effective field goal percentage is $\frac{42+4}{91} = .505$, while Los Angeles' is $\frac{39+4.5}{92} = .473$. San Antonio's shooting rating is $.505 - .473 = .032$, so San Antonio shot better. (Los Angeles' rating will be the negative of San Antonio's; in this case, $-.032$.) (2) San Antonio's turnover rate is $\frac{9}{91-10+10.4+9} = .0896$, while Los Angeles' rate is $\frac{11}{92-16+14.8+11} = .1081$. San Antonio's turnover rating is $.0896 - .1081 = -.0185$, so again San Antonio's rating is better. By the way, notice that the denominators are similar: 100.4 for San Antonio and 101.8 for Los Angeles. Since the denominator represents possessions, the numbers should be within one of each other. (3) San Antonio's offensive rebounding rate is $\frac{10}{49} = .204$ and Los Angeles' is $\frac{16}{54} = .296$. San Antonio's rebounding rating is $.204 - .296 = -.092$, so San Antonio lost the rebounding battle. (4) San Antonio's free throw rate is $\frac{19}{91} = .209$ to Los Angeles' $\frac{20}{92} = .217$. San Antonio's free throw rating is $.209 - .217 = -.008$, so San Antonio narrowly lost the free throw battle.

Each team won two of the four ratings, indicating that the game was close. In fact, San Antonio won 111-107.

Wayne Winston's book *Mathletics* notes that correlations between any two of the factors are close to zero. Mathematically, we would say that the Four Factors are "orthogonal." This means that the information contained in one factor does not duplicate the information in any other factor. This is a desirable, and somewhat rare, property.

Hockey analytics are not as advanced as baseball or basketball analytics. Hockey is very fluid with numerous line changes, making it difficult to isolate individual performances. With few goals scored, statisticians have little scoring data to work with. Hockey was an early adopter of the plus-minus statistic. The flaws of plus-minus are magnified in hockey, since goals are infrequent and are often scored on power plays. Penalty killers will have bad plus-minus numbers, unless adjustments are made for man-up situations.

An important finding in hockey is that shots taken have more predictive power than goals. That is, when predicting future success you are more likely to be right if you use the number of shots taken in past games than if you use the number of goals scored in past games. Good teams create more shots than their opponents. Bad luck (shots that barely miss or goalies that make amazing saves) can keep a good team from scoring goals, but the number of shots is a good measure of ability. This leads to two statistics, which are named for the people who popularized them.

Fenwick measures the number of shots. As just noted, this is a good predictor of future success. To be exact, start by counting shots (on goal or not). A team's **Fenwick rating** is given by shots for minus shots against. A player's Fenwick rating is shots for minus shots against while that player is on the ice. Basically, the old plus-minus statistic based on goals has been updated to the better statistic of shots. To make it a rate statistic, compute the **Fenwick percentage** equal to shots for divided by total shots. The **Corsi rating** is identical, except that blocked shots (where a defenseman blocks the puck before it reaches the net) are included in the counts.

Example 7.8 While Patrick Kane is on the ice, his Chicago team takes 12 shots, of which 3 are blocked by defensemen. The opposition takes 6 shots, 1 of which is blocked. Compute Kane's Fenwick and Corsi percentages.
Solution For Fenwick, 9 of the Chicago shots and 5 of the opponents' shots count. Kane's Fenwick rating is $9 - 5 = +4$ and his Fenwick percentage is $\frac{9}{9+5} = .643$. For the percentage, above .5 is better than average. For the Corsi rating, count all shots. Kane's Corsi percentage is $\frac{12}{12+6} = .667$.

Football (the American version) has a long history of statistics. In 2014, DeMarco Murray led the NFL in rushing with 1845 yards. Drew Brees led the NFL in passing with 4952 yards, while Andrew Luck led the league with 40 touchdown passes. These are all counting statistics. That does not make them worthless, but it probably does indicate the difficulties of isolating individual contributions in football. For example, suppose you learn that Brees just threw a 75-yard touchdown pass. You might envision a perfectly thrown long pass, but it may have been a 5-yard slant pattern that the receiver broke for a touchdown. How much credit does Brees deserve?

An early attempt at accounting for all of the credits and debits in a quarterback's performance is the NFL's Quarterback Rating Formula. An abbreviation of PR (for "Passer Rating") will be used. Like basketball's Four Factors, there are four components of the formula: completion percentage, average yards gained per attempt, touchdowns per attempt, and interceptions per attempt. Like the NBA's Four Factors, all components are rates. According to nfl.com, the following four calculations start the process for a quarterback with c completions in a attempts for y yards with t touchdowns and n interceptions. Compute (1) $(100c/a - 30) * 0.05$, (2) $(y/a - 3) * 0.25$, (3) $(100t/a) * 0.2$, (4) $2.375 - (100n/a) * 0.25$. For each of the four components, replace numbers greater than 2.375 with 2.375 and numbers less than 0 with 0. Add the four numbers (each of which is between 0 and 2.375), multiply by 100/6, and you have PR.

You may be glad to know that we can do some work to make this clearer.

Example 7.9 In 2014, Drew Brees completed 456 of 659 passes for 4952 yards, with 33 touchdowns and 17 interceptions. Compute his Quarterback Rating.
Solution The four calculations are (1) $(100 * 456/659 - 30) * 0.05 = 1.960$, (2) $(4952/659 - 3) * 0.25 = 1.129$, (3) $(100 * 33/659) * 0.2 = 1.002$ and (4) $2.375 - (100 * 17/659) * 0.25 = 1.730$. Since all four numbers are between 0 and 2.375, no adjustments are necessary. Combine the numbers to get $\text{PR} = (1.960 + 1.129 + 1.002 + 1.73) * 100/6 = 97.02$.

This rating placed Brees 6th among quarterbacks who threw at least 100 passes in 2014. (Tony Romo was first with a rating of 113.2.) The rating is not easy to compute, and harder to logically decipher. The four categories make sense, but where do those constants come from? Let's do some algebra and see what pops out. Start with

$$(100c/a - 30) * 0.05 + (y/a - 3) * 0.25 + (100t/a) * 0.2 + 2.375 - (100n/a) * 0.25$$

and multiply to get $5c/a - 1.5 + 0.25y/a - 0.75 + 20t/a + 2.375 - 25n/a$. Combine like terms to get $0.125 + \frac{5c+0.25y+20t-25n}{a}$. The Quarterback Rating can be rewritten as

$$\text{PR} = (0.125 + \frac{5c + 0.25y + 20t - 25n}{a}) * \frac{100}{6}$$

for cases in which each of the four components is between 0 and 2.375.

We can now see how the different variables are valued. One touchdown is equivalent to $20/0.25 = 80$ yards, one interception is equivalent to 100 yards, and one completion is equivalent to 20 yards. Where did these numbers come from? We will find better constants later in this chapter, and a better quarterback rating system in Chapter 10.

Evaluation and Prediction

The word "prediction" is used in different ways in this chapter. The numbers of runs a baseball team scored and allowed should predict the team's winning percentage. This means that a team that scored 665 runs and allowed 614 in 162 games should have won about 87.5 games. The games have already been played, so this is not a prediction about what will happen in the future. By contrast, a team that after 40 games has scored 181 runs and allowed 146 runs can be predicted to win 98.1 games out of 162. This is predicting what will happen in the future.

Sports analytics is interested in both types of predictions. The first type, using what happened with one statistic to estimate what happened with another statistic, is used to validate statistics to show that they have a strong relationship to what they are intended to measure. The second type is of more interest, which is to use statistics derived from past performance to predict performance in the future. Since we do not have data about the future, this second type of prediction is more difficult.

There are different ways to simulate future predictions. One way is to use data from two years ago to try to predict the results from one year ago. The resulting model could then be used to predict how many goals a given team will score next year, using this year's goals or this year's shots. A slightly different question is of a more theoretical nature and was alluded to earlier. Which statistics are most indicative of high quality play? The measure of "indicative" is an ability to predict results. We split our data in half, and use the first half of the data to create a good prediction of the second half of the data. Dividing the data chronologically can be problematic, if trades and injuries change a team dramatically in the second half of the season. You can, instead, divide the data in half by putting the odd-numbered games in one half and the even-numbered games in the other half. You still have half of the season to "predict" but it is less likely that the data will be contaminated by large changes in the team. It is in this sense that past numbers of shots in hockey predict future goals better than past goals do.

Regression to the Mean

If a player has a break-out season, performing far better than in the past, what should you predict for the future? Among the options are to project an even higher level, the same level, or a lower level of performance. A number of variables could go into this projection (age being an important one), but in general the safest bet is to predict the player to drop back closer to an average performance. This is called **regression to the mean**.

Using team passing yards from nfl.com, a linear equation for using yards p_{13} in 2013 to predict yards p_{14} in 2014 is

$$p_{14} \approx p_{13} - 0.33(p_{13} - 3770).$$

This equation is the "best" predictor in the sense that out of all equations of the form $a\,p_{13} + b$ this equation's predictions are closest to the actual 2014 values (in the "least squares" sense discussed in the next section). The significance of the constant 3770 is that it is the average passing yards for a team in 2013. So, our prediction is to take the previous year's total and remove one-third of the difference between the previous year's total and the average total. The prediction for next year is closer to the mean than last year's value was.

Example 7.10 In 2013, Denver passed for 5444 yards and Buffalo for 3103 yards. Use the regression equation to predict passing totals for Denver and Buffalo for 2014.
Solution For Denver, the equation gives us $5444 - .33(5444 - 3770) = 4892$ yards, a considerable reduction. In fact, Denver had 4661 passing yards in 2014. For Buffalo, the equation gives us $3103 - .33(3103 - 3770) = 3323$ yards, a nice increase from the below-average total in 2013. In fact, Buffalo had 3614 yards.

Regression to the mean explains a number of sports superstitions. A rookie who has a great season is subject to a "sophomore slump" that lowers production in season two. Unless the player is truly outstanding, regression to the mean explains a second season that is closer to average than the first season. Players and teams appear on the cover of *Sports Illustrated* shortly after outstanding performances. An immediate drop in production may be due to regression to the mean instead of the "SI cover jinx." A team fires its manager or coach and immediately starts performing better. Regression to the mean predicts that a bad team will improve, closer to average, whether the manager is fired or not.

Linear Weights: A Prelude to WAR

We evaluated batting average and on-base percentage using the line that best fits the data in Figure 7.3. In Example 7.10, the equation of the best fit line is used to predict passing yards for NFL teams. Both are examples of **linear regression**, which we develop more fully in this section.

Figure 7.6 shows the average score on the PGA Tour in 2013 as a function of hole length for three distances. The score increases as the distance increases, in what appears to be a linear fashion. If we had an equation for the best fit line, we could predict average scores for any distance. By "best fit" we mean, as always, that the sum of the squares of errors is minimized. Suppose the line has equation $y = mx + b$ for constants m and x. For a hole of length $x = 420$ yards, the predicted score is $m * 420 + b$ and the actual score is 4.01. The

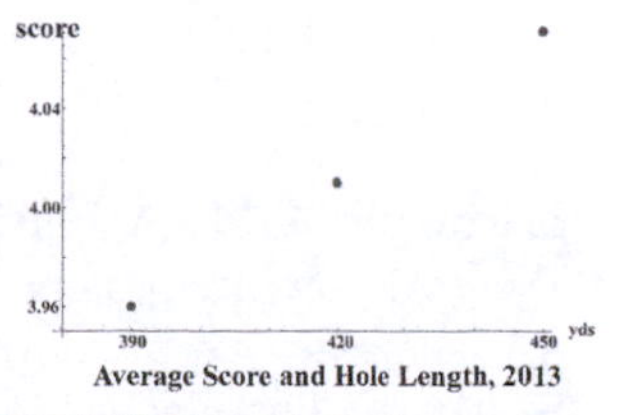

FIGURE 7.6: Score as a Function of Distance

square of the error is $(420m + b - 4.01)^2$. The calculus box below shows how to find m and b to minimize SSE. We get $a = 0.00185$ and $b = 3.24$.

The NFL Quarterback Rating combines four statistics into a single rating. The weights (the constants multiplying the variables) are mysterious, however. Ideally, the weights would be chosen to minimize the sum of the squares of errors for points scored. This is what **multiple regression** does. In this case, we want the linear combination of the four variables c/a, y/a, t/a, and n/a that minimizes SSE for team points scored. Each team has c completions in a attempts for y yards with t touchdowns and n interceptions. Using team totals for 2013 and 2014 from nfl.com, we get the following regression formula (scaled so that the coefficient of y/a is the same 0.25 as in the NFL system):

$$0.51 + 0.32(c/a) + 0.25(y/a) + 25.29(t/a) - 11.47(n/a).$$

The constants here are not totally different from those used by the NFL. The largest difference is that the NFL weights completions far too heavily (5, versus the optimal 0.32). The NFL system was created in 1973 when most passes were thrown far down field, making completions more important. The modern passing game of short passes makes completions less significant, and this shows up in the reduced weight for completions. Interestingly, touchdowns are weighted almost exactly the same and interceptions are weighted only half what they were in 1973. Despite its obscure presentation, PR (the NFL Quarterback Rating) is at heart a regression of four basic statistics against points scored.

Multiple regression is also used to create a prediction of runs scored in baseball from basic counting statistics like singles (1B), doubles (2B), and so on. The model is often called **Linear Weights**. Using values from Tom Tango's website tangotiger.net, one version of Linear Weights is given by

$$1.409(\text{HR}) + 1.063(\text{3B}) + 0.764(\text{2B}) + 0.474(\text{1B}) + 0.330(\text{BB} + \text{HBP}) + \\ 0.195(\text{SB}) - 0.456(\text{CS}) - 0.299(\text{outs}).$$

The coefficients in the Linear Weights formula give us information about the relative worth of events. We see that a walk is almost as productive as a single, giving some validation to the use of on-base percentage instead of batting average. Linear Weights is not kind to base stealers. Practitioners of "Moneyball" de-emphasize base stealing, based on calculations like the following.

Example 7.11 A player steals 14 bases in 21 attempts. Based on Linear Weight, how many runs has he created? What is the success rate needed for a base stealer to break even?
Solution The player has SB = 14 and CS = 7. The runs created equal $0.195 * 14 - 0.456 * 7 = -0.46$. The player actually cost his team about half a run. To break even, we need 0.195(SB)=0.456(CS) or $\text{CS} = \frac{195}{456}$ SB. This leads to a success rate of $\frac{\text{SB}}{\text{SB+CS}} = \frac{\text{SB}}{\text{SB+0.428SB}} = 0.700$.

By these numbers, a base stealer must be successful more than 70% of the time, or he is hurting his team! In 2014, major league totals were 2764 steals in 3799 attempts, a 72.7% success rate. On the whole, teams have learned to be smart about how often to attempt to steal bases.

Linear Weights can be used to rate players' offensive contributions. Substitute in a player's yearly totals and you get an estimate of the number of runs the player created. Since we can translate runs into wins (approximately 10 runs for one win), this allows us to estimate how many wins a player creates. This is the concept behind WAR (Wins Above Replacement), an all-encompassing statistic that attempts to measure the full range of a player's contributions (batting, fielding, baserunning, pitching) in terms of wins created compared to an average replacement player at that position.

Calculus Box: Linear Regression

The details of finding a (best fit) regression line are given here. The data points in Figure 7.6 are $(390, 3.96)$, $(420, 4.01)$, and $(450, 4.071)$. The errors for predictions from the line $y = mx + b$ are $390m + b - 3.96$, $420m + b - 4.01$, and $450m + b - 4.071$. Then

$$SSE = (390m + b - 3.96)^2 + (420m + b - 4.01)^2 + (450m + b - 4.071)^2$$

which we want to minimize. In calculus, at its minimum point a function has all first order partial derivatives either equal to zero or nonexistent. The partial derivative with respect to m gives the equation

$$2(390m+b-3.96)*390+2(420m+b-4.01)*420+2(450m+b-4.071)*450 = 0$$

and the partial derivative with respect to b gives the equation

$$2(390m + b - 3.96) + 2(420m + b - 4.01) + 2(450m + b - 4.071) = 0.$$

We solve the two equations and two unknowns through substitution, elimination, or matrix inversion. The equations are $531000m + 1260b = 5060.55$ and $1260m + 3b = 12.041$. Solving the second equation for b gives $b = 4.014 - 420m$. Substituting into the first equations gives $531000m + 1260(4.014 - 420m) = 5060.55$ or $1800m = 3.33$ and then $m = 0.00185$. Then $b = 4.014 - 420 * .00185 = 3.24$.

Roger Maris and the Hall of Fame

I grew up as a Roger Maris fan. (Mickey Mantle lived nearby and I went to high school with his sons, so I am also a fan of The Mick.) There is not a large movement to get Roger Maris into the Hall of Fame, beyond the extent to which his 61 home runs in 1961 are already memorialized. However, Tom Clavin and Danny Peary's biography of Maris makes an interesting case for his abilities as a baserunner and fielder in addition to his home run hitting.

Here are some basic Maris stats: lifetime batting average of .260 (not good), 275 home runs (good, not great, given an injury-shortened 12-year career), two MVP awards (excellent). The detailed statistics that will be discussed in Chapter 10 are not available for players from the 1960s, but play-by-play data is fairly complete going back to 1940. Using data from baseball-reference.com, a statistic called base-out runs added estimates the number of runs that a batter or baserunner adds compared to average. Maris ranks 151st for career totals, surrounded by Steve Garvey, Bernie Williams, Hall of Famer Enos Slaughter, and Ken Griffey. As a fielder, Maris was a top-five right-fielder in Range Factor in each of the years that he played right field primarily (he often filled in as a center fielder). An estimate of the total runs saved as a right-fielder has Maris ranked 29th among right-fielders.

All in all, in career WAR value Roger Maris ranks 351st, between Kirk Gibson and Chili Davis. It could be that the version of WAR used by baseball-reference.com underestimates Maris's contributions to his team, but the evidence here is for a career that was outstanding but, due to injury, falls short of Hall of Fame status.

Now Trending

Not all useful information requires detailed statistical analysis. Often, you can find interesting patterns with very simple tools. A **time series** is a sequence of values of some variable at discrete moments in time. Here, we look at some year-by-year progressions to quantify basic changes in the way sports are played.

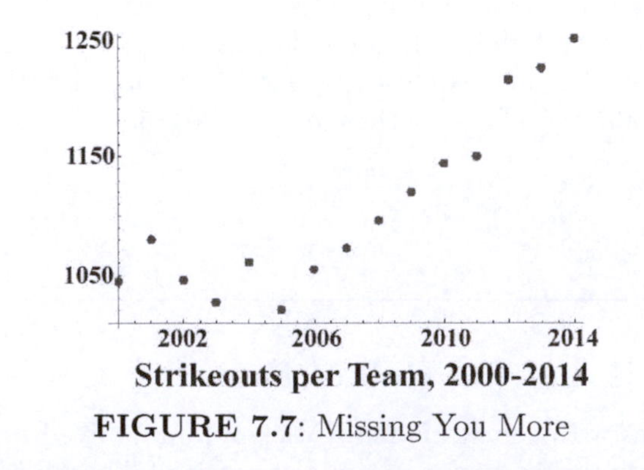

FIGURE 7.7: Missing You More

Figure 7.7 shows the alarming rate at which major league strikeouts have increased since 2005. (Data from baseball-reference.com.) The increase is almost linear. However, it would be silly to use a best fit line to predict future strikeouts. Presumably, the players or rules-makers will decide that this trend is harmful to the game and do something about it. The figure is a good test case for interpreting a time series. The ups and downs from 2000-2005 are likely a product of "random" variation, while the lengthy and steady rise from 2005 to 2014 represents a real trend that deserves attention.

Figure 7.8 shows the average passing yards per team in the NFL. (Data from pro-football-reference.com.) Passing yards has also shown a steady increase since 2005.

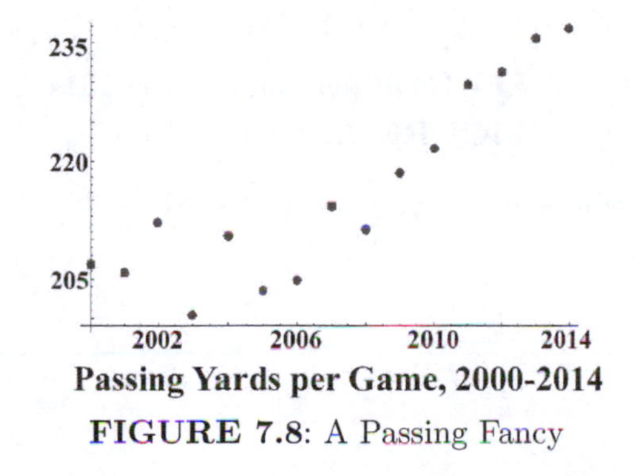

FIGURE 7.8: A Passing Fancy

Figure 7.9 shows the average number of three-point shots attempted per game. (Data from basketball-reference.com.) You can clearly see the increased usage of the shot over time since its introduction in 1979 (when a mere 2.8 shots were attempted per game). Note that the legend "2014" refers to the 2014-15 season. Analysts claim that layups and three-point shots are the most efficient shots in the game, and teams are clearly using the three-point shot more often.

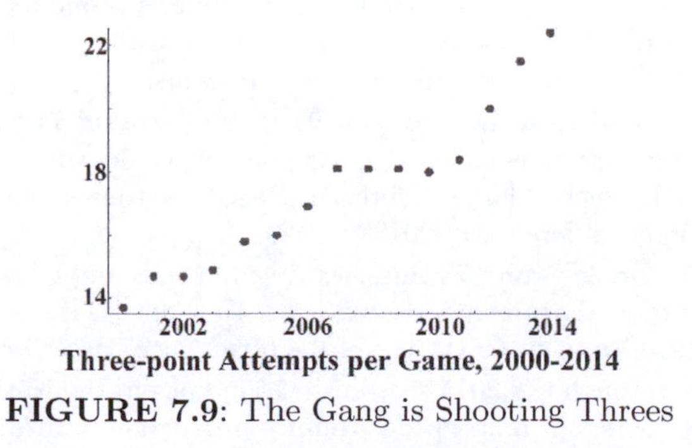

FIGURE 7.9: The Gang is Shooting Threes

Figure 7.10 shows that a trend that is "well known" is not occurring. There has been widespread discussion that scoring in soccer is diminishing, with doomsday predictions of 0-0 draws becoming the norm. Contrary to this belief, the number of goals in the English Premier League has been (until the 2014/15 season) quite high. (Data from soccer-europe.com.)

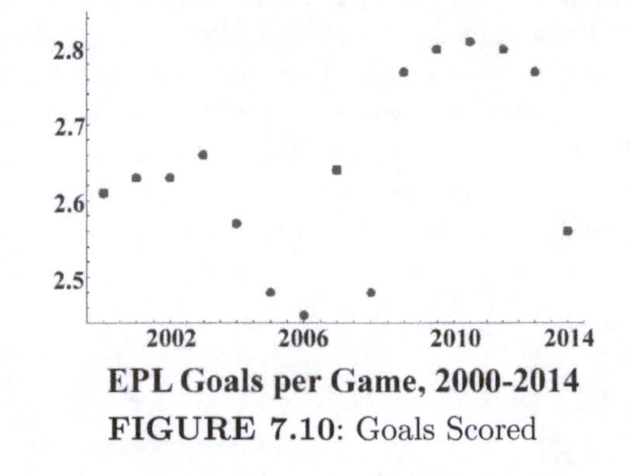

EPL Goals per Game, 2000-2014

FIGURE 7.10: Goals Scored

More trends will be explored in the exercises.

Exercises

In these exercises, (T) refers to thinking problems, conceptual problems requiring no calculations. (C) refers to problems requiring significant calculations or calculus. (P) refers to projects; these are ideas for further investigation (hints and resources are at the book's web site).

7.1 The 1983 Chicago Cubs scored 701 runs and allowed 719 runs. Use the Bill James Pythagorean Method to predict the number of wins for the Cubs over a 162-game schedule.

7.2 The 1983 Cubs won 71 games. Explain why this is one reason for Bill James to predict in his 1984 *Baseball Abstract* that the Cubs would do better in 1984. (In fact, the Cubs won their division with 96 wins.)

7.3 The 2012 Baltimore Orioles won 93 games, scoring 712 runs and allowing 705. Compare actual wins to Pythagorean expected wins. Give reasons why there might be such a large difference. Based on your explanation, would you expect a similar difference in 2013?

7.4 The 2013 Orioles won 85 games, scoring 745 runs and allowing 709. Compare actual wins to Pythagorean expected wins. How does this compare to the difference in 2012? Comment on the use of the term "luck" describing the differences.

7.5 Halfway through the 2014 season the standings in the National League West were San Francisco in first with a winning proportion of .573, then Los Angeles at .554, Colorado at .427, Arizona at .422, and San Diego at .420. Use the runs scored/against figures for each team to compute the Pythagorean winning proportions: San Francisco 338/304, Los Angeles 350/302, Colorado 405/427,

Arizona 333/400, San Diego 240/300. Which team does the Pythagorean Method predict to win the division? The season-ending winning proportions were .543, .580, .407, .395, .475. Did the Pythagorean Method correctly predict the winner? For these five teams, did the halfway proportions or the halfway Pythagorean proportions better match the final winning proportions?

7.6 Look up the MLB standings for 2014 and identify one team with a large positive "luck" value and one team with a large negative "luck" value. Track that team's record over ten years. Does that team have a predictable pattern of positive or negative luck?

7.7 On June 11, 2015, the Cleveland Cavaliers narrowly out-rebounded the Golden State Warriors 49-44. Cleveland made 29 out of 88 shots and had 16 offensive rebounds, while Golden State made 36 of 77 shots and had 6 offensive rebounds. Compute offensive rebounding rates and describe the rebounding difference.

7.8 The average number of runs scored by major league teams in 2000 was 832. (baseball-reference.com) In succeeding years, the average was 773, 747, 766, 779, 744, 787, 777, 753, 747, 710, 694, 701, 675, and 659 (in 2014). Plot the time series and discuss if there is a long-term trend for run production to decrease.

7.9 Look up the average number of walks per team from 2000-2014, plot the time series, and discuss any trends. Repeat for home runs and hits.

7.10 The average number of points scored by NFL teams in 2000 was 20.7. (pro-football-reference.com) In succeeding years, the average was 20.2, 21.7, 20.8, 21.5, 20.6, 20.7, 21.7, 22.0, 21.5, 22.0, 22.2, 22.8, 23.4, and 22.6 (in 2014). Plot the time series and discuss if there is a long-term trend for points to increase.

7.11 Look up the average number of rushing yards per team from 2000-2014, plot the time series, and discuss any trends. Repeat for first downs.

7.12 The pace (average number of possessions per team) in the NBA teams in 2000-01 was 91.3. (basketball-reference.com) In succeeding years, the average was 90.7, 91.0, 90.1, 90.9, 90.5, 91.9, 92.4, 91.7, 92.7, 92.1, 91.3, 92.0, 93.9, and 93.9 (in 2014-15). Plot the time series and discuss if there is a long-term trend for the pace to increase.

7.13 Look up the average field goal percentage per team from 2000-2014, plot the time series, and discuss any trends. Repeat for free throw percentage.

7.14 Look up the offensive and defensive ranks of teams in championship games, use the ranks to label the team with the better offense and better defense, and compare how often the better offense wins versus how often the better defense wins. (a) NFL 1970-2014; (b) NBA 1970-2014; (c) NHL 1970-2014.

7.15 In 2014, Gold Glove-winning shortstop J.J. Hardy made 13 errors in 594 chances. Compute his fielding percentage. Find how many chances another fielder could have handled to have the same fielding percentage with 12 errors. Which shortstop would be more valuable?

7.16 In 1965, Hall of Fame shortstop Luis Aparicio made 20 errors in 697 chances. Compute Aparicio's fielding percentage and compare to J.J. Hardy's in 2014. Given that both played in 141 games (and both for Baltimore), which shortstop do you think had the better year? What other information would be good to have?

7.17 In 2014-15, the Golden State Warriors played the lineup Curry-Thompson-Barnes-Green-Bogut for 812 minutes, outscoring the opponents by 358 points.

The lineup Curry-Thompson-Barnes-Green-Speights played for 201 minutes, outscoring the opponents by 78 points. Compute the net points per minute for each lineup, and discuss the relative worth of Bogut and Speights. What other information would you want to know to make a better judgment?

7.18 In 2014, the home and away home run numbers for 81 games each are (a) Colorado: $H_h = 119$, $R_h = 67$, $H_a = 90$, $R_a = 83$; (b) Boston: $H_h = 49$, $R_h = 74$, $H_a = 67$, $R_a = 87$; (c) San Diego: $H_h = 54$, $R_h = 55$, $H_a = 47$, $R_a = 70$; (d) Baltimore: $H_h = 107$, $R_h = 104$, $H_a = 68$, $R_a = 83$. Compute home run park factors for each team.

7.19 In 2014, the home and away hits numbers for 81 games each are (a) Colorado: $H_h = 924$, $R_h = 627$, $H_a = 825$, $R_a = 703$; (b) Boston: $H_h = 696$, $R_h = 659$, $H_a = 760$, $R_a = 698$; (c) San Diego: $H_h = 597$, $R_h = 602$, $H_a = 598$, $R_a = 702$; (d) Baltimore: $H_h = 705$, $R_h = 729$, $H_a = 670$, $R_a = 672$. Compute hits park factors for each team.

7.20 In 2013, Miguel Cabrera was chosen as MVP over Mike Trout. Their stats are listed. Compute Linear Weights for each. Also, compute a similar stat called **weighted on-base percentage** (wOBA) given by (.72 BB + .75 HBP + .90 1B + 1.24 2B + 1.56 3B + 1.95 HR)/PA. Discuss your results.

	BB	HBP	1B	2B	3B	HR	PA	SB	CS	outs
Trout	110	9	115	39	9	27	716	33	7	399
Cabrera	90	5	122	26	1	44	652	3	0	362

7.21 Compute the Four Factors and discuss the fact that the Warriors won both games.

(a)

	FGA	FGM	3P	FTM	OREB	DREB	TO
Warriors	88	39	10	20	11	37	12
Cavaliers	94	39	9	13	13	32	11

(b)

	FGA	FGM	3P	FTM	OREB	DREB	TO
Warriors	85	37	13	18	7	32	9
Cavaliers	82	32	6	27	16	40	16

7.22 In 2014-15, Alex Ovechkin had 217 shots on goal, 109 missed shots, and 141 blocked shots in 1215 minutes. Compute Ovechkin's Fenwick, Corsi, and Fenwick per 60 minutes ratings. Repeat for Patrick Kane, who had 124 shots on goal, 42 missed shots, and 54 blocked shots in 908 minutes. (Data from hockeyanalysis.com.)

7.23 In 1966, Don Meredith completed 177 passes in 344 attempts for 2805 yards with 24 touchdowns and 12 interceptions. Sonny Jurgensen completed 254 passes in 436 attempts for 3209 yards with 28 touchdowns and 19 interceptions. (a) Compute the NFL Quarterback Rating PR for each. (b) The NFL totals were 3149 completions in 6108 attempts for 37436 yards with 280 touchdowns and 318 interceptions. In 2014, the NFL totals were 11200 completions in 17879 attempts for 121247 yards with 807 touchdowns and 450 interceptions. Compute the league PR ratings for the two years.

7.24 On 11/27/14, Tony Romo completed 18 passes in 29 attempts for 199 yards with 0 touchdowns and 2 interceptions. On 12/21/14, he completed 18 passes in 20 attempts for 218 yards with 4 touchdowns and 0 interceptions. Compute his PR for each game.

7.25 Determine the maximum PR for a game. Give an example of statistics that achieve the maximum.

7.26 The best fit line for predicting 2014 rushing yards from 2013 rushing yards is approximately $r_{14} = 1383 + .22r_{13}$. Show that this gives (approximately) the regression to the mean equation $r_{14} = r_{13} - .78(r_{13} - 1780)$, where the average rushing yards for a team in 2013 was 1780.

7.27 The best fit line for predicting 2014 rushing yards per attempt from 2013 rushing yards per attempt is approximately $ra_{14} = 3.44 + .169ra_{13}$. Show that this gives (approximately) the regression to the mean equation $ra_{14} = ra_{13} - .831(ra_{13} - 4.2)$, where the average rushing yards per attempt for a team in 2013 was 4.2.

7.28 The regression to the mean equation has the form $y = x - c(x - a)$ where x is the previous value, a is the average value and c is a constant. Show that if $c = 1$, the prediction y equals the average value; if $c = 0$, the prediction y is the previous value. Based on this, does a persistent statistic correspond to a larger or smaller value of c?

7.29 (T) Give several reasons why a team's actual record could deviate from its Pythagorean Method prediction. Discuss whether that trend should continue or not; that is, are you describing a repeatable skill or luck?

7.30 (T) Figure 7.3 indicates that on-base percentage was more important for teams in 2004 than batting average. Assuming that teams acted on this information, what would be the impact in terms of playing time for players who do or do not walk often? Explain why this might produce the neutral situation of Figure 7.5.

7.31 (T) A small budget team like the Oakland A's needs to find undervalued players. Use Figures 7.3 and 7.5 to explain why the A's could find high-OBP players for cheap in 2004 but not in 2014. Comment on the "moving target" nature of finding undervalued players.

7.32 (T) Describe a situation in which a counting statistic is a better representation of the value of a player than the corresponding rate statistic, and a situation in which the rate statistic is better.

7.33 (T) Discuss the importance of a statistic being persistent.

7.34 (T) In golf, Strokes Gained putting is not especially persistent (correlation of 0.44). Give at least two possible reasons having to do with the skill of putting.

7.35 (T) If a personal statistic is not persistent for players who change teams or coaches, discuss whether this statistic measures a skill or not.

7.36 (T) Give at least two possible reasons why winning percentage for a pitcher is less persistent than strikeouts.

7.37 (T) Give two advantages of Range Factor over fielding percentage in evaluating a fielder. Do you agree with the idea that fielding percentage made more sense back in the day when gloves were very small?

7.38 (T) Discuss the extent to which the plus-minus statistic depends on when and how often a player plays in a game.

7.39 (T) *Basketball Reference* defines "true shooting percentage" (TSP) as points divided by 2(FGA+.44FTA). Compared to effective field goal percentage (EFG), compare how TSP and EFG handle three-point shots, missed shots, and free throws.

7.40 (T) When approximating the number of possessions by a team, explain why offensive rebounds are subtracted.

7.41 (T) Explain why a team's defensive rebounds would not necessarily equal its opponent's field goals missed.

7.42 (T) Discuss why the NFL Quarterback Rating limits each of its four components to being between 0 and 2.375.

7.43 (T) Explain why a rating with coefficients calibrated to accurately estimate certain values (e.g., points) in one year might not accurately predict the same values for the next year.

7.44 (T) Describe a situation in which a team's statistics for the first half of the season might not accurately predict its statistics for the second half of the season. Explain why determining coefficients using odd-numbered games and testing on even-numbered games can be better than determining coefficients with the first half of the season and testing on the second half of the season.

7.45 (T) A coach praises a player, and the player's performance declines; the coach yells at a player, and the player's performance improves. Explain this phenomenon with regression to the mean. Discuss the psychological implications of this.

7.46 (T) (a) After a golfer shoots a personal best score, what would you predict happens the next round? Explain. (b) After a mediocre team gets a great upset victory, what would you predict happens the next game? Explain.

7.47 (T) The coefficient for pass completion percentage is much smaller in our regression than in the NFL Quarterback Rating formula. Compared to 2015, in 1973 teams threw very few short passes. Explain why completions are less valuable in 2015 than they were in 1973.

7.48 (C) Look up runs scored and allowed for all MLB teams in 2014. Find the exponent (to two digits) that minimizes the sum of the squares of the errors in Pythagorean Method predictions.

7.49 (C) Repeat exercise 7.48 for (a) 2014 NFL; (b) 2014-15 NBA. How do the exponents relate to the average number of points scored in a game?

7.50 (C) Repeat exercise 7.48 for 2014-15 (a) NHL; (b) EPL. Decide whether you want to predict wins (a draw is half a win) or points. Does the Pythagorean Method seem more or less accurate for hockey and soccer compared to baseball?

7.51 (C) Stanley Rothman's linear equation for NFL wins is $0.5 + .001538(\text{PF} - \text{PA})$ for a team that scores PF points and allows PA. For the 2014 season, compute the sums of squares of errors for the Pythagorean Method with exponent 2.5 and for the Rothman method. Which method performs better?

7.52 (C) In this exercise, we use calculus to derive Rothman's equation from the Pythagorean Method. Think of the Pythagorean formula as giving wins as a function of runs scored x and runs allowed y: $w(x,y) = \frac{x^2}{x^2+y^2}$. The linear approximation of this function is $L(x,y) = w(a,b) + w_x(a,b) * (x-a) + w_y(a,b) * (y-b)$ where w_x and w_y are the partial derivatives (to be explained) and a and b are typical values of x and y, respectively. Take $a = b = 725$, an average number of runs scored by a team in 162 games. The first term in the linear approximation is $w(725, 725) = 0.5$ as in Rothman's equation. Next, compute w_x, which is the derivative of w treating x as the variable and y as a constant. This is a quotient rule and $w_x = \frac{2xy^2}{(x^2+y^2)^2}$. We get $w_x(725, 725) = \frac{1}{1450} \approx 0.0007$. Show that $w_y(725, 725) = -\frac{1}{1450}$ so our linear equation is $w(x,y) \approx 0.5 + 0.0007(x - 725) -$

$0.0007(y-725) = 0.5+0.0007(x-y)$ which is a rounded off version of Rothman's equation.

7.53 Ⓒ Find the linear approximation for the more general $w(x,y) = \frac{x^a}{x^a+y^a}$ and compare to Rothman's equations $0.5 + .001538(PF - PA)$ for the NFL, $0.5 + .000351(PF - PA)$ for the NBA, and $0.5 + .002102(GF - GA)$ for the NHL.

7.54 Ⓒ Create separate scatter plots for each MLB statistic in 2004 and 2014 (in each case, use runs scored for y) and compare, commenting on any differences. (a) RBI; (b) HR; (c) SLG.

7.55 Ⓒ Create scatter plots for each NFL statistic in 2014 (in each case, use points scored for y) and comment on which statistic is the "best." (a) rushing yards; (b) passing yards; (c) rushing yards per attempt; (d) passing yards per attempt.

7.56 Ⓒ Create scatter plots for each NBA statistic in 2014 (in each case, use points scored for y) and comment on which statistic is the "best." (a) field goals made; (b) field goal percentage; (c) rebounds; (d) offensive rebound percentage.

7.57 Ⓒ Compute the persistence (autocorrelation) for each statistic for teams in 2013 and 2014. (a) Football yards rushing; (b) Football yards passing; (c) Basketball field goal percentage; (d) Basketball free throw percentage.

7.58 Ⓒ For NBA teams in 2014-15, construct scatter plots and compute correlations using points as one variable with the other variable being (a) FTM/FGA; (b) FTA/possessions. (Estimate possessions using FGA−OREB+0.4FTA+TO.) Discuss the use of FTM/FGA in the Four Factors.

7.59 Ⓒ Compute the average NFL Quarterback Rating for (a) 1973, (b) 1983, (c) 1993, (d) 2003, and (e) 2013. Discuss how the game is changing.

7.60 Ⓒ For the English Premier League in 2013/14, find the best linear equation for team goals in terms of shots, possession time, passing percentage, and fouls. (Data can be found at whoscored.com.) Discuss the influence of each variable.

7.61 Ⓒ Find the best fit line for the data (1,1), (2,4), (3,9). Explain why it is silly to fit a line to this data. Given this example, discuss the importance of graphing your data before finding a regression equation.

7.62 Ⓟ Track the errors in the Pythagorean Method for some sport. Compute the autocorrelations for consecutive years. (The correlation of both lists, one of which has the luck values for each team in one year and the other of which has the luck for the teams in the same order the next year.) Does "luck" seem to be the right word for the errors?

7.63 Ⓟ Taking pace of play into account, study whether the better offensive team in a championship game/series wins more often than the better defensive team wins.

7.64 Ⓟ The Four Factors are considered to be "orthogonal" so that the correlation between any two of them should be near zero. Use team season totals to explore whether the factors are orthogonal.

7.65 Ⓟ Find the value of k that best predicts runs scored by a team in a season with the formula k OBP + SLG. How does OPS ($k = 1$)

7.66 Ⓟ For your favorite sport, compute the "winning percentage" for a variety of statistics. For example, in football you could look at TD passes; find the percentage of games in which the team with more TD passes wins the game.

Further Reading

The book *Moneyball* is by Michael Lewis, who also helped with the screenplay for the movie. *Big Data Baseball* is a more recent description of the Pittsburgh Pirates' conversion to analytics and their rise to prominence. *The Only Rule Is It Has to Work* describes a season using analytics to run a minor league team.

The Sloan Sports Analytics Conference baseball panels (videos available online) have much more information about Oakland's scouting practices. Nate Silver's excellent book *The Signal and the Noise* also mentions Oakland's scouting budget.

A derivation of the general Pythagorean Method is given in Steven J. Miller's "A derivation of the Pythagorean Won-Loss Formula in Baseball."

The Music of Pythagoras is an enjoyable review of what is known and not known about the historical Pythagoras and his followers. *The Cult of Pythagoras* takes a more skeptical look at the evidence.

The 1982-88 *Bill James Baseball Abstracts* were published by Ballantine Books, and are enjoyable reads even today. *Hardball Times* and *Baseball Prospectus* publish baseball annuals that are in the same vein that James mined.

Annual guides include *The Fielding Bible* from Baseball Info Systems, the *Football Outsiders Almanac*, Hardball Times' *Baseball Annual*, and guides from the Prospectus family (Baseball, Basketball, and Hockey).

The Numbers Game: Why Everything You Know About Soccer is Wrong discusses scoring trends in various soccer leagues. A summary can be found at the soccerbythenumbers.com web site. *Soccermatics* is a nice complement, exploring other forms of analytics.

Excellent books about sports analytics include *Baseball Between the Numbers* by Baseball Prospectus, *The Sabermetric Revolution* by Baumer and Zimbalist, *Mathletics* by Winston, *Basketball on Paper* by Oliver, *Basketball Analytics Spatial Tracking* by Shea, *Analytic Methods in Sports* by Severini, *Analyzing Wimbledon* by Klaasen and Magnus, *Stumbling on Wins* by Berri and Schmidt, and *The Book* by Tango and Lichtman.

Chapter 8

Randomness in Sports

Introduction

The lead pass to Marvey'o Otey was thrown too far. Otey, playing for William Byrd High School in Vinton, Virginia, on December 9, 2013, chased the ball down but only had enough time to get his right hand on the ball and whip it behind his back. Then Otey's momentum carried him out of bounds and through an open door out of the gym and into a school hallway. Hearing the crowd cheering, Otey thought his desperate save must have been grabbed by a teammate. Only later did he learn that his save had gone in the basket for one of the most outrageous three-point baskets ever.

FIGURE 8.1: Behind the Back and Straight Out the Door

There is no doubt that luck plays a role in sports. Otey's shot, the football pass that ricochets off of three defenders right into a receiver's hand for a touchdown, the line drive that bounces off the wall at an odd angle: these pieces of good luck are balanced by the shot that beats the goalie only to bounce off of a post, or the putt that hits the wrong blade of grass and veers away from the hole.

The extent to which luck affects results is open to debate, and is the subject of this chapter. As we explore this general issue, the following questions will be addressed. Which sport is most subject to chance? Does the best team always win? Which sports leagues have the most balance? Does the "hot hand" exist? Do balanced scoring teams win the NBA championship? Are base hits a matter of luck?

Summing Up the Basics

Some basic probability and statistics tools are needed to follow many analytics discussions, whether they be about sports or politics or business. Mean, standard deviation, and distributions are briefly introduced here. To make the discussion more concrete, we use the two sets of numbers g = {27, 34, 22, 21, 30, 28} and r = {17, 12, 31, 18, 9, 15}. These are the points scored in the first six games of the second round of the 2015 NBA playoffs by Blake Griffin and J.J. Redick, respectively.

To see which player is higher scoring, you can compute the means. The **mean** is what most people think of as "the average" even though other versions of averages (like the median) are preferred by statisticians. The mean of g is $\frac{27+34+22+21+30+28}{6} = 27$ and the mean of r is $\frac{17+12+31+18+9+15}{6} = 17$. Griffin is higher scoring.

Both players showed several ups and downs in their scoring patterns, but Griffin was much more consistent than Redick. We measure consistency with variance and standard deviation. To compute the **variance**, usually denoted s^2, subtract the mean from each value, square the result, add and divide by one less than the number of values $(6 - 1 = 5)$. For g, we first compute {0, 7, −5, −6, 3, 1}, then {0, 49, 25, 36, 9, 1}, then the sum 120 and finally the variance $\frac{120}{5} = 24$. For r, we get a variance of $\frac{290}{5} = 58$.

The **standard deviation** is simply the square root of the variance, and is denoted by s. (We denote the variance by s^2 so that the standard deviation has a simple representation. This is an indication that the standard deviation will be more useful to us than the variance.) For Griffin, we get $s = \sqrt{24} = 4.9$ and for Redick $s = \sqrt{58} = 7.6$. This quantifies the fact that Griffin's points were more consistent.

The **empirical rule** of statistics states that for bell-shaped data, about 68% of the points will be within one standard deviation of the mean, 95% within two standard deviations of the mean, and 99.7% within three standard deviations of the mean. We have no information about whether or not an individual player's point totals are bell-shaped, but let's see how this might work. For Blake Griffin, within one standard deviation of the mean gives the range 27 ± 4.9, or between 22 and 32; in fact, 4 out of 6 (67%) of his point totals are in this range. Plus or minus two standard deviations is 27 ± 9.8, or from 17 to 37; all of his point totals are in this range. For Redick, two-thirds of the point totals are in the range 17 ± 7.6, or between 10 and 24. Two standard deviations gives the range 17 ± 15.2, or between 2 and 32; this contains all of his values. Notice that with Redick's inconsistency a much wider range of values is needed to capture his actual output.

The empirical rule assumes a bell shape for the data. The *distribution* of data is often overlooked in basic analyses. The bell curve of the normal distribution is a common occurrence, but it should be checked before using the empirical rule.

Figure 8.2 is a **histogram** or bar graph of points scored by the Los Angeles Clippers during the 2014-15 NBA regular season. Each bar represents a point range of four points, and is drawn to a height representing the number of times data points in that range occurred. For example, the tallest bar shows that there were 16 games in which the Clippers scored between 104 and 108 points. The bars do not form a perfect bell curve, but there is a clear peak in the middle with a nearly symmetric drop-off to each side.

FIGURE 8.2: Histogram

Figure 8.3 overlays a bell curve on top of the histogram of Figure 8.2. (The equation for this curve will be explored in the exercises.) With this visual, we can see that the histogram is approximated reasonably well by a bell curve. This is enough to justify using the empirical rule. Even an 82-game season is not long enough for us to demand that the distribution form a perfect bell curve.

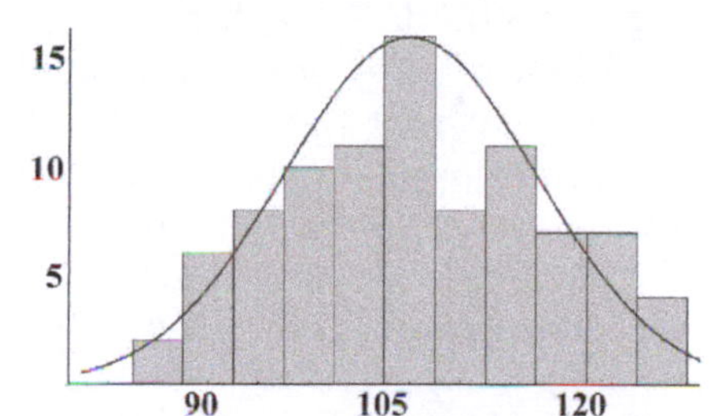

FIGURE 8.3: Normal Curve

There are numerous distributions of importance that are *not* normal. A course in probability will introduce several common distributions. To illustrate a different distribution that occurs with regularity, Figure 8.4 shows a histogram for points per game in 2014-15 for all NBA players who averaged at least 2 points per game. The first bar from the left shows that 82 NBA players averaged between 2 and 5 points per game in 2014-15. As the scoring average increases, the number of players with that average decreases at a regular rate. This is an example of a *power law* distribution (to be explored in exercise 8.49).

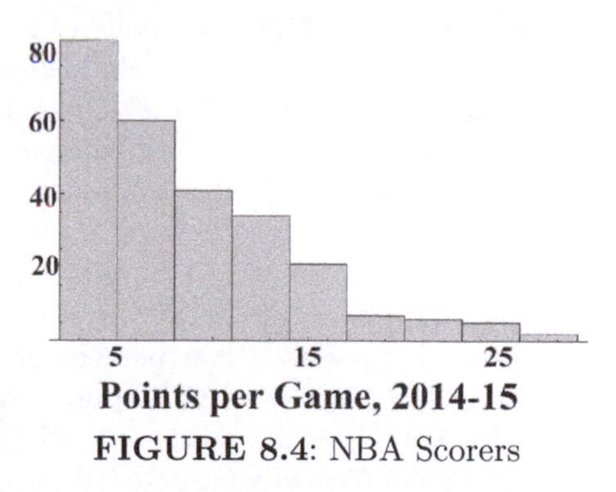

FIGURE 8.4: NBA Scorers

Prediction is Difficult

The physicist Niels Bohr is credited with saying, "Prediction is very difficult, especially about the future." It is not recorded whether or not he had just visited Las Vegas. Most sports bets are "against the spread" so that the bet feels fair. You may have no doubt that San Antonio will defeat New York, but if the spread is 15 points then San Antonio must win by more than 15 points for you to win a bet on San Antonio.

To illustrate how risky such a bet is, suppose that your friend is a big New York fan, and is willing to bet on New York with a spread of 3 points. The experts who determine the spread are saying that San Antonio is 15 points better, but the Spurs need to win by only 4 points or more for you to win your bet. That sounds like a great bet! But, how likely are you to win? Take a guess: 80%? 90%? higher?

To answer the question, we need to know the distribution of NBA scores. It turns out that the difference between the spread and the actual outcome is (approximately) normally distributed with mean 0 and standard deviation 12. On average, then, the spread gives the outcome: San Antonio by 15. Using the empirical rule, we know that about 68% of the results will be within one standard deviation of the mean. So, 68% of the results will have San Antonio winning by between $15 - 12 = 3$ and $15 + 12 = 27$ points. More importantly, the other 32% of the time the result will not be between 3 and 27, with 16% less than 3 and 16% greater than 27. You will lose your bet 16% of the time! It's true that having an 84% chance of winning a bet is good, but this is probably not as high a probability as you were expecting.

Wayne Winston's book *Mathletics* gives standard deviations for scores in several leagues. The NBA is 12 points, and college basketball is 10 points; I would have guessed that the younger college players would have more fluctuations in their scores. The standard deviation for NFL scores is 14 points; for college football scores it is 16 points. Even though the total points scored is lower, scores in football vary more than in basketball!

Suppose that you have a system that consistently beats the NFL point spread by 4 points. What percentage of bets will you win? Take a guess, but keep in mind that the NFL standard deviation is 14 points. Calculus can be used to compute that you should win a little more than 61% of the time. If you want to win 65% of your bets, you need to increase your margin of superiority over the spread to 5.4 points.

It is not easy to have a high winning percentage in Las Vegas!

A Slump or a Disaster

Two main goals of analytics are prediction and evaluation. We have seen that randomness can make prediction difficult. The same is true of evaluation. Suppose that a baseball team opens the season with 10 wins and 15 losses. Is this a bad team? If the team has a lot of talent, is it time to fire the manager?

A baseball team with a winning percentage of 60% will win 97 games over a 162-game season. This is more than enough wins to make the playoffs. Is it possible that such a team could have a 10-15 start just by bad luck? The question is asking us to compute the probability that a 60%-quality team could have a 10-15 start. To answer this question, we start by making assumptions, giving us a model with which probabilities can be computed.

The simplest model is a **binomial model** under which games are **Bernoulli trials**. The terminology is important because it gives us a shorthand to describe a common set of assumptions. In particular, a game has two possible outcomes (win or lose), and we assume that our team has a 60% chance of winning each game. That means that the games are **independent**: the outcomes of past games do not affect the probability of the next game. We are thus ignoring the effects of good and bad pitching opponents (every game has the same win probability) as well as the effects of good and bad streaks (games are independent). We do not have to believe that this is true, but it allows us to compute a probability and draw conclusions.

We can evaluate the probability in multiple ways. We start by calculating mean and standard deviation. To do so, we use the following facts about binomial distributions. If x is the number of times a particular outcome (e.g., a win) of a sequence of n Bernoulli trials of probability p occurs, then the mean of x is np, and the standard deviation for x is $\sqrt{np(1-p)}$. The following example shows how this works.

Example 8.1 If a baseball team wins games with probability $p = 0.6$ and the games are independent, compute the mean and standard deviation of the number of wins in 25 games.

Solution The main assumptions of a binomial distribution are present: two outcomes (win or lose) per try, constant probability, and independence. We have probability $p = 0.6$ and $n = 25$ trials. The mean is $np = 25 \cdot 0.6 = 15$. This should make sense: 60% of 25 is 15. The standard deviation is not as common sensible, so it is nice to have the formula $\sigma = \sqrt{np(1-p)} = \sqrt{25 \cdot 0.6 \cdot 0.4} = \sqrt{6} \approx 2.45$.

The empirical rule applies for binomial distributions with large values of n. We expect the number of wins to be within two standard deviations of the mean 95% of the time. In this case, we look at $15 \pm 2 \cdot 2.45$ or 10.1 to 19.9. Our team's 10-win total is right at the edge of this. What do we conclude? *If* the games were completely random, a bad streak worse than 10-15 would occur about 2.5% (half of 5%) of the time, or about one in 40. Purely on the basis of the team's record, there is reason to stay patient: bad streaks can occur by chance. Of course, the games are not completely random. If the 25 games were all at home against bad teams and the manager was alienating players and fans, then perhaps a change of manager is in order. However, keep in mind regression to the mean, which says that this team is likely to bounce back to its 60%-win mean.

Calculus Box: Probability

The probability in Example 8.1 can be computed directly. Let's start with a smaller example. What is the probability that the 60% team starts with 1 win and 3 losses? For example, the team could win its first game and then lose three in a row. The game-by-game probabilities of this happening are 0.6, 0.4, 0.4, and 0.4. Because the games are assumed to be independent events, we can multiply these probabilities together to get a probability of $(0.6)(0.4)^3$ for the sequence WLLL. This is not the only way to start 1-3. Notice that the sequence LWLL would generate the same probability: $0.6 \cdot 0.4 \cdot 0.4 \cdot 0.4 = 0.4 \cdot 0.6 \cdot 0.4 \cdot 0.4 = (0.6)(0.4)^3$. So, the total probability of starting 1-3 will equal $(0.6)(0.4)^3$ times the number of different orderings for one W and three L's. We have listed WLLL and LWLL so far; how many more are there? One way to answer this is to focus on the one win: there are four games in which that win could occur, so there are four different orders. The probability is $4(0.6)(0.4)^3$.

The general structure of our answer can be used for all problems of this type. If there are W wins and L losses, then

Binomial probability = (Number of orders) $* \; p^W * (1-p)^L$.

All that remains is to find a simple way to count the number of sequences. This is done using a formula called the **binomial coefficient**. For our calculation, we want to know how many different ways 10 wins can be arranged in a sequence of 25 games. The binomial coefficient for this is named "Twenty-five choose ten" and has several notations: $\binom{25}{10}$, ${}_{25}C_{10}$, and C(25,10) are common ones. The formula uses the **factorial** which is defined as $n! = n(n-1)(n-2)\cdots(2)(1)$. For example, $5! = 5 \cdot 4 \cdot 3 \cdot 2 \cdot 1 = 120$. Then the binomial coefficient equals

$$\binom{n}{m} = \frac{n!}{m!(n-m)!} = \frac{n(n-1)(n-2)\cdots(n-m+1)}{m(m-1)(m-2)\cdots 1}$$

where the second version of the formula emphasizes that the same number of factors are in the numerator and denominator. Before tackling the numbers in Example 8.1, a simple calculation shows that $\binom{6}{3} = \frac{6\cdot5\cdot4}{3\cdot2\cdot1} = 20$. This is the number of different orders of wins and losses with 3 wins in 6 games.

Example 8.2 If a baseball team wins games with probability $p = 0.6$ and the games are independent, compute (a) the probability of winning exactly 10 games out of 25; (b) the probability of winning at most 10 games out of 25.
Solution The assumptions of the binomial distribution are met. (a) The probability of winning 10 games and losing 15 is given by $\binom{25}{10}(.6)^{10}(.4)^{15}$ with $\binom{25}{10} = \frac{25!}{10!15!}$. Putting this all together, the probability is $\frac{25!}{10!15!}(.6)^{10}(.4)^{15} \approx 0.0212$. (b) To the value in part (a), we need to add the probability of winning 9 games, $\frac{25!}{9!16!}(.6)^{9}(.4)^{16}$, and the probability of winning 8 games, $\frac{25!}{8!17!}(.6)^{8}(.4)^{17}$, and so on. The sum can be written as

$$\sum_{n=0}^{n=10} \frac{25!}{n!(25-n)!}(.6)^{n}(.4)^{25-n}$$

and equals approximately 0.0344.

You may have noticed that the value in Example 8.2 is larger than that of Example 8.1. This is a result of applying the empirical rule when the number of games is not large. The issue has to do with the empirical rule applying to continuous distributions in which the probability of a single outcome is essentially zero. There is no distinction between the probability of the value of the variable being less than 10 versus being less than *or equal* to 10. In Example 8.2, the probability of the number of wins being less than or equal to 10 is 0.0344, but the probability of wins being less than 10 is 0.0132. The average of the two is 0.0238, which is very close to that given by the empirical rule.

May the Best Team Win

A playoff-caliber team can have a slump over a 25-game stretch. With a 162-game schedule, it seems likely that the best baseball team will prevail in the long run. A quick calculation puts that in doubt, and a simulation run by Bill James gives surprising results.

Example 8.3 If a baseball team wins games with probability $p = 0.6$ and the games are independent, compute the mean and standard deviation of wins over a 162-game season. If it takes 91 wins to make the playoffs, how likely is the team to do so?
Solution The assumptions of the binomial distribution are met. The mean is $(0.6)162 = 97.2$ or about 97 wins. The standard deviation is $\sqrt{(162)(.6)(.4)} =$

6.2 or about 6 wins. Since 91 wins is one standard deviation away, the empirical rule gives the team a 16% chance of having less than 91 wins, or an 84% chance of making the playoffs. (The calculus box above shows how to get the exact value, which is about 86%.)

The *1989 Baseball Abstract* includes an essay by Bill James on this issue. James simulated 1000 seasons of the then-current league of 26 teams in 4 divisions. He first randomly assigned each team a "true quality" based on the historical distribution of winning percentages in baseball. So, a team's quality was chosen randomly from a normal distribution with mean 0.5 and standard deviation 0.05. The computer then simulated every game for a season, using the official schedules. This raises a significant question: if team A of quality 0.55 plays team B of quality 0.52, how often should team A win? The answer should logically be greater than 0.5 (team A is better) and less than 0.55 (team B is above average). James used the conditional probability formula $\frac{0.55(1-0.52)}{0.55(1-0.52)+(1-0.55)0.52} = 0.530$, which is now known as the **log5 method**.

The results of the simulation are surprising. The best team (greatest true quality) in a division won the division 54.6% of the time - barely over half! The best team in baseball failed to win its division 28.5% of the time, and won the simulated World Series 29.3% of the time. That is, the best team was as likely to miss the playoffs as it was to win it all.

A problem that James had with his simulation helps explain the surprising result. The actual spread of wins and losses in the simulations produced a larger standard deviation than the value that he used to create the true qualities. This is due to the number of wins for one team not really being independent of the number of wins for another team. If an underdog team A gets an unexpected ("lucky") win over division rival team B, then team B also just got an unexpected loss. A small number of upsets between division rivals can dramatically change the standings since division games count double (a win for one plus a loss for the other).

Measuring Parity: Gini in a Bottle

Which of the major sports leagues has the most parity? Leagues such as the NFL enforce salary caps and arrange the draft of new talent to favor the teams with the worst records. Does this actually result in parity? In the first 49 years of the Super Bowl, 19 of the 32 teams (57%) won at least once while 28 (87.5%) made the Super Bowl at least once. In the last 49 major league baseball seasons, 20 of the 30 teams (66.7%) won at least one World Series and 27 of the 30 teams (90%) made it to the Series at least once. Baseball, in spite of a limited salary cap and decreased importance of the draft, shows more parity. In the last 49 years of the NBA, 15 out of 30 teams (50%) won the title while 21 out of 30 (70%) made the finals. There seems to be less parity in basketball.

There are ways to measure parity other than simply counting up champions and runners-up. The Gini index is used by economists to measure inequality of wealth. Imagine a league with the ultimate parity: each of 100 teams wins exactly half of its games, with 5 wins and 5 losses. Take the bottom 5% of the teams (the five worst teams) and add up their wins. We have 5 teams with 25 wins, which is exactly 5% of the 500 total wins in the league. The bottom 10% of the league (10 teams) will have 50 total wins, exactly 10% of the total wins in the league. The **Lorenz curve** of percentile versus percentage of wins would include the points (5,5) and (10,10) based on our discussion so far. The full graph would be the straight line shown in Figure 8.5.

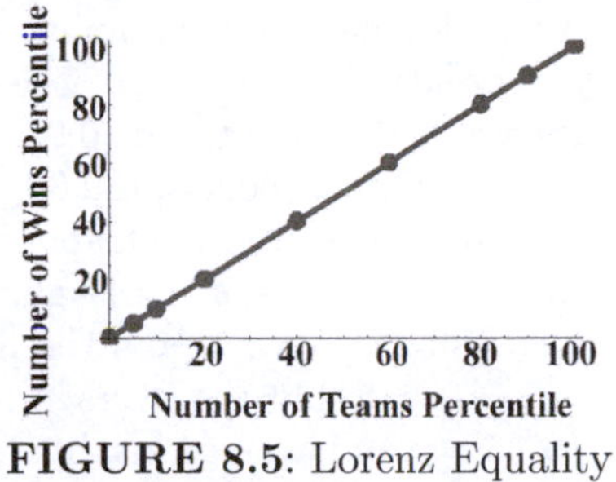

FIGURE 8.5: Lorenz Equality

Now, imagine the same league with a large disparity, with the five worst teams each losing all 10 games and the next five worst teams winning twice each. The total number of wins for the bottom 5% is 0, so the Lorenz curve includes the point (5,0). The bottom 10% of the league won a total of 10 games, which is 2% of the 500 total wins. The Lorenz curve includes the point (10,2). If the next 10 teams each won 3 games, then the bottom 20% of the league won $10 + 30 = 40$ games, which is 8% of the total: the Lorenz curve includes the point (20,8). The full graph might look like Figure 8.6.

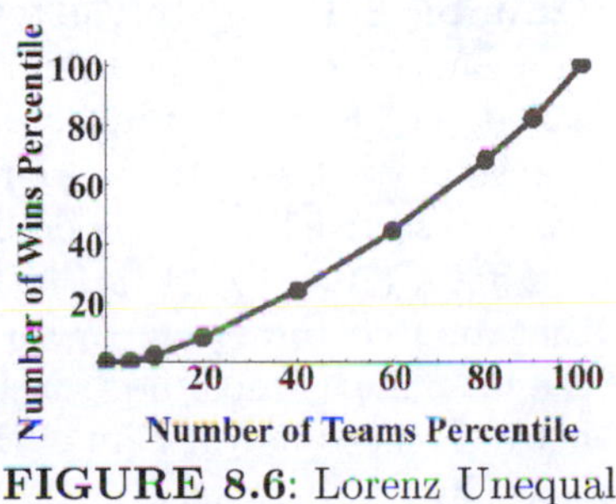

FIGURE 8.6: Lorenz Unequal

The more inequality there is in the league, the farther the Lorenz curve will get from the total-equality curve of Figure 8.5. So, the gap between the Lorenz curves in Figure 8.5 and 8.6 is a measure of the disparity in the league. The **Gini index** is a ratio of areas that quantifies this statement. The denominator is the area between the ideal curve (Figure 8.5) and the x-axis. This is a triangle of width 100 and height 100, which has area 5000. The numerator is the area between the actual Lorenz curve and the ideal curve, as shown in Figures 8.7 and 8.8.

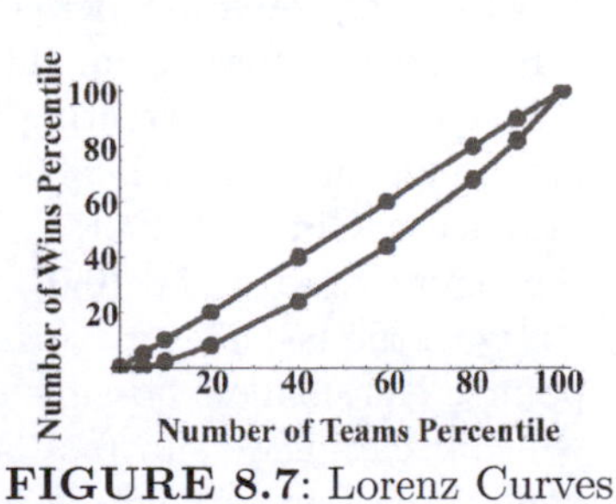

FIGURE 8.7: Lorenz Curves

The calculation of the area for regions such as the one shown in Figure 8.8 can require calculus. An estimate can be made from the data by summing up areas of trapezoids. Suppose that the data points shown in Figure 8.6 are (0,0), (5,0), (10,2), (20,8), (40,24), (60,44), (80,68), (90,82), and (100,100). The area of a trapezoid is the base times the average of the heights. For the region using the x-axis and the first two data points (0,0) and (5,0), the width is $5 - 0 = 5$ and the average height is $\frac{0+0}{2} = 0$. The next region has width $10-5 = 5$ and average height $\frac{0+2}{2} = 1$. The third region has width $20-10 = 10$ and average height $\frac{2+8}{2} = 5$. The area under the Lorenz curve is given by $5\cdot 0+5\cdot 1+10\cdot 5+20\cdot 16+20\cdot 34+20\cdot 56+10\cdot 75+10\cdot 91 = 3835$. The area *between* the curves is the difference $5000 - 3835 = 1165$ and the Gini index is $\dfrac{1165}{5000} = 0.233$.

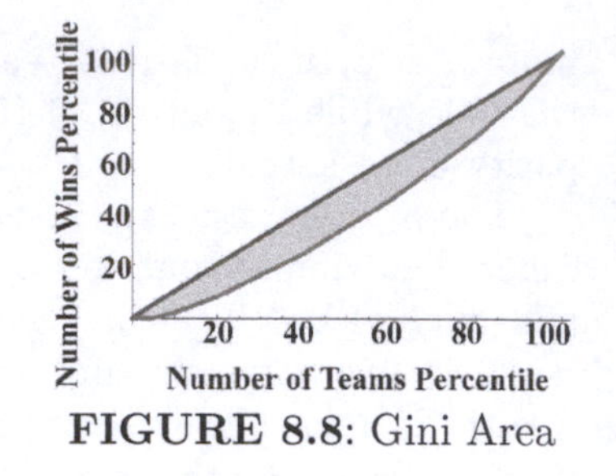

FIGURE 8.8: Gini Area

Example 8.4 A basketball team has players score 34, 28, 26, 8, and 4 points in a game. Plot the Lorenz curve and calculate the Gini index.
Solution The total points scored by the team is 100. The bottom 20% (the lowest of five) scored 4% of the points. The bottom 40% (the two lowest scorers) scored 12% of the points, and so on. The points for the Lorenz curve are (0,0), (20,4), (40,12), (60,38), (80,66), and (100,100). The Lorenz curve that connects these points with line segments is shown in Figure 8.9. As before, the denominator for the Gini index is 5000. The numerator is the difference between 5000 and the sum of the areas of trapezoids. The trapezoid areas are given by width times average height. We have $20\frac{0+4}{2} + 20\frac{4+12}{2} + 20\frac{12+38}{2} + 20\frac{38+66}{2} + 20\frac{66+100}{2} = 3400$. The Gini index equals $\frac{5000-3400}{5000} = 0.32$.

The Gini index is always between 0 and 1, with 0 representing perfect equality and 1 representing total inequality. In Example 8.4, since we are subtracting from 5000, the maximum numerator is 5000, which would produce a Gini index of 1. To subtract 0, the points need to be (0,0), (20,0), and so on, with the last player scoring 100% of the points. The smallest possible numerator is 0, which occurs if we subtract 5000. This happens if the Lorenz curve is $y = x$, occurring if every player scores the same number of points.

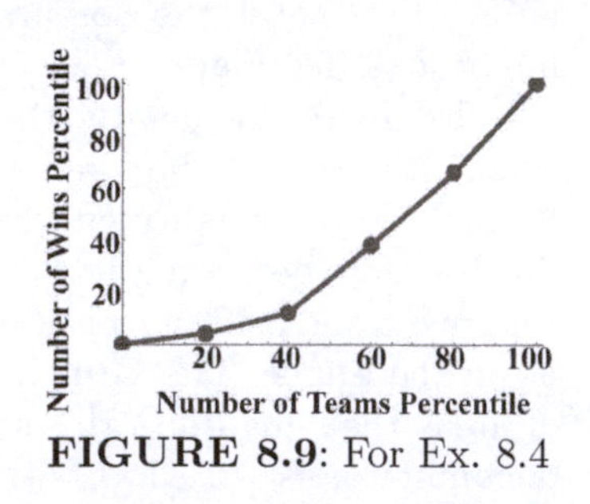

FIGURE 8.9: For Ex. 8.4

Gini indices for 2014 are shown in Table 8.1. By this measure, then, baseball has the most equality. This should make sense, given that there are often NBA and NFL teams that win 80% of their games and the best baseball team barely wins 60% of its games.

TABLE 8.1: Gini Indices

NBA	0.18
MLB	0.06
NFL	0.17

Measuring Parity: Luck versus Skill

A different way to measure league balance can be thought of as measuring the percentage of luck in the sport's games. This terminology is a little misleading, but the idea is to look at win variance in a league compared to a league whose games are completely random. The 2014-15 NBA season gives us a test case.

Imagine two alternate-universe NBAs, one in which every game is determined by a coin flip and one in which the better team always wins. Call these the Luck-NBA and the Skill-NBA. We can compute the variance of wins for each league. The Luck-NBA has a binomial distribution with $n = 82$ games and $p = .5$ for each game. The variance for wins of an individual team is $np(1-p) = 20.5$. In the Skill-NBA, the best team goes 82-0, the second-best team goes 80-2 or 79-3 or 78-4 depending on how many times the teams played (determined by which divisions the teams are in). Using a value of 3 games per team, the variance of the number of wins is about 620. The variance for the number of wins in the real NBA in 2014-15 was about 181. The real variance is between the Skill-NBA variance and the Luck-NBA variance. Our goal is to determine the right combination of skill and luck to produce the actual variance.

The model we use is that a game result is a fraction p of skill and a fraction $1-p$ of luck. If $p = .4$, then the result is 40% skill and 60% luck. In an equation, we assume that

$$r = p\,s + (1-p)\,k$$

where the luck k has mean 0 and is independent of skill. With these assumptions, the result variance equals p^2 times the skill variance plus $(1-p)^2$ times the luck variance. For the NBA, we want

$$181 = p^2(620) + (1-p)^2(20.5)$$

which we can solve to get $p = 0.53$. We should not conclude that NBA games are half luck, but we can use this number to see how balanced the NBA is compared to other leagues. The larger the value of p, the more often the better team wins, and the less balanced the league is. The p value for MLB wins in 2014 is $p = 0.16$. Certainly, the better team wins less often in baseball than basketball. Wins in the 2014 NFL have $p = 0.61$, the largest of the three. By this measure, then, the NFL has the least balance, and the better team wins the most often.

The Paradox of Skill

A thought experiment takes this in a different direction. Suppose that all players have the same high skill level. Then every contest would be determined by a gust of wind, bounce of the ball, or referee's call; in other words, luck. This is the **paradox of skill**: the less variation in skill there is, the more important luck is. Now, think of the skill levels of players being described by a bell-shaped curve. Most players are average, but a few are especially good and a few are especially bad. Figure 8.10 shows a bell curve with a vertical line drawn to the right to indicate a maximum human level of performance.

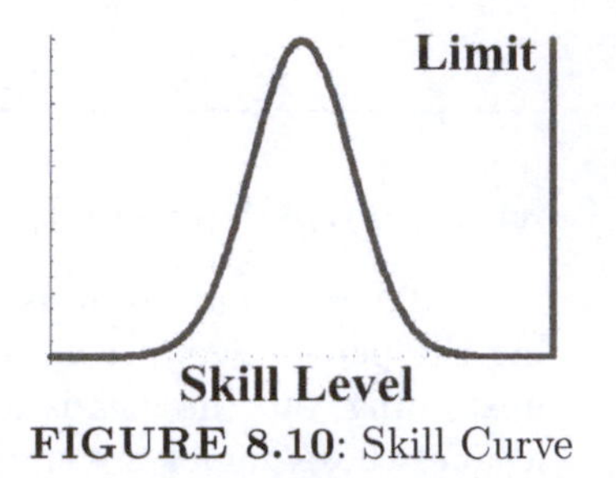

FIGURE 8.10: Skill Curve

If all players get better, what happens? At first, the curve in Figure 8.10 could simply move to the right. However, as the distribution begins to approach the limit of human performance, the distribution will necessarily pile up near the limit. Figure 8.11 shows the beginning of this process, with the obvious side effect of the spread of the distribution decreasing.

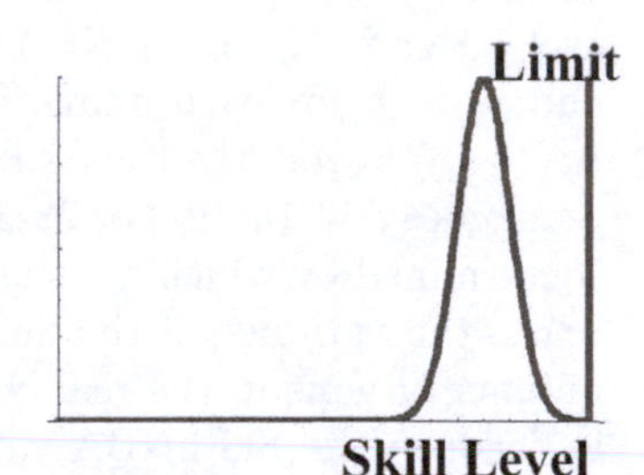

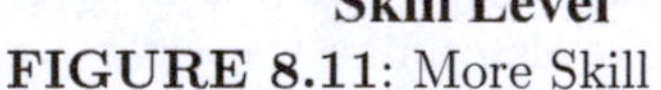

FIGURE 8.11: More Skill

Stephen Jay Gould used this idea to explain why no baseball player has had a season batting average over .400 since Ted Williams in 1941. In baseball, pitchers and rule makers combine to keep league batting averages from increasing. A league-wide increase in baseball skill would not result in increased batting averages. By the above argument, the result could be a reduced standard deviation, which would reduce the odds of someone having a remarkably high batting average. To see why, consider two Boston Red Sox hitters, Ted Williams batting .406 in 1941 and Wade Boggs batting .368 in 1985. Which is more impressive? In *Keep Your Eye on the Ball*, Watts and Bahill compute league averages of .282 in 1941 and .268 in 1985, and standard deviations of .0340 in 1941 and .0264 (smaller!) in 1985. Williams's average is 3.65 standard deviations above the league average, and Boggs's average is 3.81 standard deviations above the league average. Boggs's .368 average in 1985 is more out of the ordinary than Williams's .406 average in 1941, due to 1941's higher league average and higher standard deviation.

As athletes get better, it becomes harder to be clearly superior to all others. To put this in terms of the paradox of skill, the better the competition is the more you need luck (and lots of it) to dominate.

Measuring Parity: Entropy

A third way to measure league balance is to adapt the concept of entropy. A cornerstone of information theory, entropy (or Shannon entropy, named for Claude Shannon) measures the amount of information in a signal in the sense that the more predictable a message is, the less information is gained when the message arrives. Maximum entropy occurs when all possible signals are equally likely (so that the signal is as unpredictable as possible). In a sports context, entropy will be maximized when every team or player has an equally likely chance to win or score; in other words, when there is maximum parity. If there are n possible outcomes which occur with probability $p_1, p_2, \ldots,$ and p_n then **entropy** is defined by

$$\text{entropy} = -p_1\ln(p_1) - p_2\ln(p_2)\cdots - p_n\ln(p_n)$$

where "ln" is the natural logarithm.

Example 8.5 The eight highest scorers in points per game in the playoffs leading up to the 2015 NBA Finals were c = {27.6, 18.7, 13.5, 10.1, 9.4, 9.1, 7.0, 4.8} for Cleveland and g = {29.2, 19.7, 14.0, 11.3, 8.0, 5.3, 5.0, 4.9} for Golden State. (Data from basketball-reference.com.) Compute the entropy for each team.

Solution For Golden State, the total points for the eight players is 97.4, so divide each value in g by 97.4 to get the proportion of points gp = {.300, .202, .144, .116, .082, .054, .051, .050}. For Cleveland, the total points for the eight players is 100.2, so the proportion of points scored is essentially equal to the values in c divided by 100. The entropy for Cleveland is
$-.276\ln(.276) - .187\ln(.187) - .135\ln(.135) - .101\ln(.101) - .094\ln(.094) - .091\ln(.091) - .07\ln(.07) - .048\ln(.048) = 1.94.$
By a similar calculation, the entropy for Golden State is 1.88.

For comparison, the maximum entropy for eight scorers is 2.08 (ln8), and a team for which one player scored 99.3 percent of the points would have entropy 0.05. (Technically, ln(0) is undefined so a one-person team would have undefined entropy, but this shows that the limiting entropy value for complete inequality is zero.) The calculation shows that Cleveland was more balanced in its scoring than Golden State.

In the 2015 playoffs, the team with the higher scoring entropy in the regular season (top ten scorers) won just 5 of the 15 playoff series. The NBA is a league of stars.

A similar calculation can be made for the entropy of wins in a league. Table 8.2 shows the entropy values for 2014 in the middle column. The entropy for baseball is higher than for basketball, indicating again that baseball has more parity. The NFL entropy is almost as high as the MLB value, but that is misleading since the NFL has more teams. The right-most column shows the league entropy divided by maximum entropy. We can see that baseball has, by far, the most parity while the NBA has more parity than the NFL.

TABLE 8.2: League Entropy

	entropy	pct of max
NBA	3.34	0.984
MLB	3.39	0.998
NFL	3.38	0.975

Declaration of Independence

In several examples, we have assumed that events such as games are independent processes. This makes the calculations easier, but we should worry about how unrealistic the assumption is. In *Analyzing Wimbledon*, Klaasen and Magnus conclude that points in a tennis game are won in an independent-like pattern. That is, if the server wins a point on serve 57% of the time, calculations of games won, deuces won, and so on, using a binomial model with $p = 0.57$ match the actual results reasonably well.

For the 2014 NBA Finals, the NBA switched from a 2-3-2 scheduling of home games to a 2-2-1-1-1 schedule. In 2013 the first two games were in Miami, the next three in San Antonio, and the last two in Miami. In 2014, the first two games were in San Antonio, the next two in Miami, and the fifth game in San Antonio, with the sixth game scheduled in Miami and the seventh game scheduled in San Antonio. If the games are independent, the sequence does not matter: the first four and the first six games are split equally in both scenarios, and the independence assumption means that order does not affect the probabilities (more on this in the exercises).

To check for independence, we can look at a variety of situations. In the NBA playoffs from 2003 to 2015, the home team won 65% of the games. If games are independent, this percentage should carry through all games. However, the home team won game one 72% of the time, game two 75% of the time, game three 56% of the time, and game four 54% of the time. There is an explanation for this difference. Games one and two are played at the home court of the better team, games three and four at the home of the lesser team. We can claim independence as long as we assign different probabilities for the two teams: 75% for the better team and 55% for the worse team seem reasonable.

Let's take a closer look at game five. The independence assumption would imply that the home team wins 75%, very close to the actual 78% figure. Independence looks good so far! The home team in game five could be ahead

3 wins to 1, or tied 2 wins to 2, or behind 1 win to 3. In 3-1 situations, the home team won 76% of the games; in 2-2 situations, the home team won 75%. However, in 1-3 situations (where another loss would eliminate the team from the playoffs) the home team won 27 out of 31 times, or 87%! To see if this is statistically significant, we can compute the probability that a 75% team would randomly win 27 or more games out of 31: the binomial model gives an 8% chance, so this does not quite qualify as statistically significant. But it is suggestive that the independence assumption does not apply to a team facing elimination!

Conditional Probability

The calculations done above are examples of **conditional probabilities**. Instead of looking at how often the home team wins game five, we look at how often the home team wins game five *given that the series is tied 2-2* or *given that the team trails 1-3*. The probability could be changed by the condition that is imposed (e.g., game five with the series tied).

A formal definition of independence is that events A and B are independent if the conditional probability of event A given that event B occurs equals the (unconditional) probability of A. That is, A has the same probability whether or not B occurs. In our NBA playoff situation, if the outcome of game five is independent of the current won-lost standings, then the probability of the home team winning is the same whether it is ahead 3-1, tied 2-2, or behind 1-3. As noted, there is some evidence that this might not be true.

When we say that game outcomes are independent, or the outcomes of at bats or free throws or whatever are independent, we are really saying that they are independent of everything. The probability of success is not changed by anything (the current score, the loud fan in the second row, the current alignment of Jupiter and Mars, *anything*). The challenge for analysts is to actively test for factors that could change the probability. Otherwise, the analyst's calculations could be meaningless.

A silly, but (unfortunately) common, example of conditional probability is the habit of announcers to give situational statistics. Bob Uecker "played" baseball before his excellent announcing and acting career, and in true Uecker style managed to finish his career with a batting average of .200. However, he batted .300 against Hall-of-Famer Steve Carlton and .333 against Hall-of-Famer Warren Spahn! Before you get excited, notice how suspiciously round those averages are. In fact, Uecker was 2-for-6 against Spahn and 3-for-10 against Carlton. The numbers are not large enough to be significant: for example, a .200 hitter would randomly get 2 or more hits in 6 at bats more than one-third of the time. (You can find data like this at baseball-reference.com.)

A better example, although only slightly better, involves Pete Rose. Rose's first hit was off of Bob Friend, against whom Rose was a robust 16-for-36. His last hit was off of Greg Minton, an excellent pitcher whom Rose touched for 13-out-of-30. As you will see in exercise 8.60, even 30 or 36 at bats is not enough to draw much of a conclusion.

The Hot Hands

All athletes know the feeling of being "in the zone" or "on fire." The target looks twice as large as normal, time moves slowly, distractions disappear, and success seems inevitable. The widespread knowledge of the sensation of the "hot hand" made it especially galling when two psychiatrists, Amos Tversky and Thomas Gilovich, published research in the 1980s claiming that the hot hand is merely a cognitive illusion. Their work centered on a simple test of independence. If a person gets hot, the probability of success increases. If successes are independent, then there is no such thing as the hot hand.

Tversky and Gilovich recorded all field goal attempts by Philadelphia 76er players in home games in the 1980-81 season. They looked at the percentage of shots made after strings of makes and strings of misses. If all shots are independent, the percentages should be equal. If hot hands exists, the percentages following strings of made shots should be higher than those following strings of misses. However, they were basically the same. They looked at free throws by Boston Celtics players during the 1980-81 and 1981-82 seasons in the same way. Again, they found no significant differences in shooting percentages. In fact, most players shot slightly better after a *miss.*

Numerous studies have followed for other sports, and the findings have been consistently negative. Streaks do not seem to occur in professional sports beyond what you would expect from a random sequence. It is important to note that the probability p of success is not always 0.5. A 90% free throw shooter will have long streaks of made free throws. The streaks just are not significantly longer than what you would get from a coin that was biased to come up heads 90% of the time. Few violations of independence have been found.

The cognitive illusion aspect of Tversky-Gilovich is a flaw of the human brain that has been demonstrated numerous times. Try this quiz: one of the three sequences of Hs and Ts was created using a coin flip model where each flip is independent. The other two sequences were created with processes that are not independent. (Of course, any of the three sequences could result from coin flipping. Randomness can be so random.)

THTTTTHTTTTHHTTHHHTHTHHTHHHTHH
HTTTHTHHTHTTHHHTHTHTTTHTHHTHTH
HTHHHHHHTTTTTTTTHHTHHHHHHTTTTT

The third sequence should look too streaky to be from coin flipping; in fact, it was created so that the previous outcome would be repeated 80% of the time. Do you think the first or second sequence shows independence? The first sequence has an early streak of 8-out-of-9 Ts; the second sequence has a nice balance of Hs and Ts in every subsequence of nine symbols. And that is what is fake about it: the rule was to do coin flipping until three in a row occurs, but never allow four in a row to occur. The third sequence is streakier than real coin flipping, while the second sequence is less streaky than real coin flipping. And yet, most people choose the second sequence as the real one, and if asked to generate a "random" sequence of Hs and Ts most people produce something like the second sequence.

The lesson is that the human brain is wired to find and explain patterns, and we have a tendency to assign meaning (like "hot" or "cold") to sequences that are actually random.

Not So Fast, My Friend

The false lead about the first sequence above ("an early streak of 8-out-of-9 Ts") leads to an important critique of much of the research into the hot hands. Making 8 out of 9 probably qualifies as hot even though it may not have an especially long streak of consecutive makes. Depending on the situation, 7 out of 9 might be noteworthy. Research that only defines hotness in terms of consecutive successes may be missing the point.

Larkey, Smith, and Kadane suggest a different definition of hotness. A basketball player who scores 10 points in 2 minutes will be thought of as being "on fire" even if he or she has missed a couple of shots. The clustering of multiple successes into short periods of time is another way to run hot and cold, even if the percentage of successes remains relatively constant. The Larkey study of the NBA found several examples (notably Vinnie "The Microwave" Johnson) of players who are streaky.

In the book *Curve Ball*, Albert and Bennett find evidence that the batting averages of some batters fluctuate significantly over the course of a season. This is not what most people mean by the hot hand, but it is another example of the binomial model not always working. In *Mathletics*, Wayne Winston notes that when the at bats are adjusted for park factors and pitching matchups, the appearance of streakiness disappears.

Bocskocsky, Ezekowitz, and Stein collected data from the 2012-13 NBA season and found evidence of several aspects of the hot hand. Using optical tracking data, they found average success rates for shots based on player ability, distance, angle, closeness of defenders, shot clock, and other variables. They then monitored each player's hotness by comparing recent success rates to the average success rates. Thus, a player who had made two out of three 3-

pointers with a hand in the face would be hotter than someone who had made four wide open layups in a row. By their measure of hotness, hot players are more likely to (1) take the team's next shot, (2) take a harder shot, (3) be more closely guarded, and (4) make the next shot. This shows that (1) teammates defer to the hot player, (2) the player is "feeling it" and gets overconfident, (3) the defense adjusts to stop the hot player, and (4) the player is shooting better than normal. Note that (2) cancels the effects of (4), which explains the inability of researchers to find long sequences of successes.

Runs Tests

Other than taking my word for which of the three sequences in the "Hot hands" section came from a coin flipping model, how can you tell? A statistical test called the **Wald-Wolfowitz Runs Test** helps us evaluate sequences of successes and failures. Define a **run** to be a sequence of the same letter. For example, SSFSFFF has 4 runs: two Ss, one F, one S, three Fs. The Runs Test uses the number of runs to analyze a sequence. Suppose a sequence (of at bats, or field goal attempts, or wins/losses) of length n includes s successes and f failures. For large n, the number of runs is approximately normal with mean μ and standard deviation σ where

$$\mu = \frac{2sf}{n} + 1 \text{ and } \sigma^2 = \frac{(\mu-1)(\mu-2)}{n-1}$$

and the empirical rule is used to evaluate the likelihood that the sequence comes from an independent and identically distributed process.

Each of the sequences above has length $n = 30$, $s = 15$ Hs, and $f = 15$ Ts. Then $\mu = 16$ and $\sigma^2 = \frac{210}{29}$ so that $\sigma \approx 2.7$. Any number of runs outside the interval $\mu \pm 2\sigma = 16 \pm 5.4$ is suspicious. Thus, 10 or less and 22 or more are suspect. The first sequence has 16 runs, exactly equal to the mean. The second sequence has 21 runs, which is almost statistically significant. The third sequence has 8 runs, which is three standard deviations below the mean and therefore significantly low.

The Runs Test does not identify the second sequence as significantly different from independence. Looking at the lengths of the runs can accomplish this. The first sequence has 16 runs. Half of them should be of length 1, half of the remainder of length 2, and so on. We expect 8 runs of length 1, 4 runs of length 2, 2 runs of length 3, 1 run of length 4, and 1 run of length greater than 4. The first sequence has 8 runs of length 1, 4 runs of length 2, 2 runs of length 3, and 2 of length 4. The second sequence has 14 runs of length 1, 3 runs of length 2, and 3 runs of length 3. This is not a good match. A chi-square goodness of fit test quantifies the mismatch.

Another way to test a sequence is to create a large number of random permutations and collect statistics. For example, create a random sequence of 15 Hs and 15 Ts. Do this a million times. How many of the sequences do *not* have a streak at least four long? My simulation had 160,217. This is evidence that the probability of a sequence having that characteristic of the second sequence above (no runs of length 4 or more) is about 16%; i.e., not likely.

Joltin Joe and The Streak

In 1941, Joe DiMaggio got hits in 56 consecutive games. This broke the old consecutive game streak of 44 games (it has since been revised to 45). The large gap between best and second-best is one reason that DiMaggio's record is revered. Stephen Jay Gould wrote, "Thus Joe DiMaggio's 56-game hitting streak is both the greatest factual achievement in the history of baseball and a principal icon of American mythology." Equipped with some statistical knowledge and play-by-play data from baseball-reference.com, let's see how impressive the record is.

During the streak, DiMaggio batted .408 (91 out of 223). If he batted 4 times in a game, the probability that he made 4 outs is $.592^4 = .123$ so the probability that he got a hit in a game is .877. He gets a hit in 56 straight games with probability $.877^{56} = .00064$ or one time in 1600. Wow! That is very unlikely. It is also a very bad analysis. What is wrong? The probability of getting a hit in a game is wrong (see below), the calculation ignores the fact that he had more than one chance to start the streak, and the calculation implicitly assumes that his at bats were independent. Let's do a little better.

The value of .877 for getting a hit in a game is too high. We used his batting average, which ignores walks, so we assumed that he had four at bats in every game without ever walking. During the streak, DiMaggio actually had 246 plate appearances, so he got a hit in 37% of his plate appearances. Using this figure in place of .408, his probability of getting a hit in a game drops to 84%. (In fact, in 1941 he got a hit in 82% of his games. In 1940, he got a hit in 86% of his games.)

A quick tangent: on May 15, 1941, both DiMaggio and Ted Williams started the longest hit streak of their major league careers. Even though Williams hit .406 for the entire 1941 season, his streak was only 23 games long. Why did DiMaggio and not Williams have the longest hitting streak? In the game that ended Williams's streak, he walked three times. By contrast, DiMaggio walked three times *in the last thirty games* of his streak in situations in which he had not yet achieved his hit. Williams, by the way, has the major league record for most consecutive games reaching base, a staggering 84 straight games in 1949.

DiMaggio did not like to walk, and during the streak the pitchers were under pressure to "play fair" and pitch to him. In addition, DiMaggio rarely struck out. During the streak, he struck out only five times, with no strikeouts in the last 32 games of the streak! This means that in almost every plate appearance DiMaggio hit a fair ball. This makes DiMaggio an excellent candidate for a long hitting streak.

Back to estimating the probability of DiMaggio's streak happening: we need to ask the question more precisely. That is, if you want the probability of hitting in those exact 56 games, then $.84^{56} \approx .000057$ is reasonable (and is impressively small). However, if you want the probability that he would have the streak *at some point* in 1941, multiply by 25 (the number of games in which he did not get a hit in 1941, and therefore the number of chances he had to start a new streak). We're now up to 0.0014, but still well less than one percent. If you want the probability that DiMaggio would have such a streak at some point in his career, multiply by 5 or so; the probability is still less than one percent. (Such probabilities will be explored in the exercises.) Numerous estimates have been made for the probability that somebody sometime in the history of baseball would have a 56-game hitting streak; two percent is a common choice.

As Stephen Jay Gould said, the streak is a great achievement. However, consider two more facts. After going hitless in game 57 (thanks to two great defensive plays by the opposing third baseman, Ken Keltner), DiMaggio had hits in the next 16 games to make it 72 out of 73! And, in 1933 DiMaggio set a minor league record by hitting in 61 straight games!

Not Following the Rules

All of the calculations in the previous section assumed that DiMaggio's at bats were independent events. This cannot be true, but how much does the assumption affect the calculations? Trent McCotter explored this question in a paper with the provocative title of *Hitting Streaks Don't Obey Your Rules*. McCotter simulated 50 years of baseball using the actual averages and schedules of the players.

The simulations gave interesting results. On the average, McCotter's simulations produced (over the 50 simulated years) 49 players with streaks of at least 25 games. In the real 50 baseball seasons, 62 players had streaks of 25 games or more. The table shows the simulation averages and actual numbers for other lengths of streaks.

TABLE 8.3: Simulated vs Actual Streaks

length	sim avg	actual
25+	49	62
30+	10	19
35+	2	5

Twice as many players as predicted by the independence model are achieving long hitting streaks. This doesn't prove anything, but it does might make you question calculations based on the independence assumption. Incidentally, note how few players reach 35 games in a row, and recall that Joe DiMaggio's streak was 56 games!

Here is, to my mind, the most convincing evidence. In the 50 years, 4 real baseball players had streaks of length exactly 29 games. McCotter's simulations produced 4 or more streaks of exactly 29 games about 25% of the time. Nothing significant here. However, 9 real ballplayers had streaks of exactly 30 games. This many 30-game streaks never happened in McCotter's 1000 simulations.

Real baseball players had longer streaks than their simulated (i.e., independent) counterparts, but they were especially likely to extend a 29-game hitting streak to 30. This looks like the result of human psychological (and non-independent) effort. A related fact involves batting averages. From 1950-2014, 89 players had a season batting average of .296, 104 players had an average of .297, 82 players had an average of .298, 60 players had an average of .299, and 178 players had a batting average of .300!

BABIP and DIPS

A baseball pitcher who gives up several hits in a row and is replaced by another pitcher is said to have been "knocked out of the box." Sometimes, the phrase seems too harsh, as hits can be swinging bunts, ground balls that sneak between fielders, and soft fly balls that barely clear the infield. Voros McCracken found a way to quantify the bad luck that dogs some pitchers. He found that approximately 30% of fair balls in play (not home runs) fall for hits. Pitchers who give up hits on more than 30% of balls in play are likely having a spell of bad luck that will disappear soon.

This leads to the **BABIP (batting average on balls in play)** statistic, computed as $\frac{\mathrm{H}-\mathrm{HR}}{\mathrm{AB}-\mathrm{K}-\mathrm{HR}+\mathrm{SF}+\mathrm{SH}}$ for a player with H hits in AB at bats, HR home runs, K strikeouts, SF sacrifice flies, and SH sacrifice hits.

Research on BABIP indicates that the main control a pitcher has over his BABIP against is whether batters tend to hit fly balls or ground balls against him. Presumably, average velocity on batted balls (data that has become available recently) will also be found to affect BABIP against. Similarly, batters have control over BABIP through percentage of ground balls, line drives, and fly balls, as well as average velocity of batted ball.

Defense also plays an important role in BABIP, as a well-positioned or outstanding individual defense will turn hits into outs. This insight leads to **DIPS**, or Defense-Independent Pitching Statistics. McCracken's version uses strikeouts, walks, hit batsmen, and home runs allowed: the aspects of the game that a pitcher can actually control. (Catchers do have a large effect on how umpires call balls and strikes, so perhaps this also needs to be modified.) The formula is a little messy because McCracken wanted it to be on the same scale as earned runs allowed, but a simplified version is $3.2 + (13\text{HR} + 3\text{W} + 3\text{HBP} - 2\text{K})/\text{IP}$.

Random Thoughts

In his book *Mathletics*, Wayne Winston tells of a statistical analysis gone bad. Historically, when college basketball games have been fixed the players are not asked to lose, but to win by less than the spread. Suppose the spread is 15 points. Then the favorite could win by 10, make the gamblers happy, and still win the game. Statistically, we might expect the results of honest games to form a bell curve with mean 15 and standard deviation 10. About 34% of the results would fall between 5 and 15 points (the empirical rule) and about 42% of the results would fall between 1 and 14 points (inclusive). By symmetry, about 42% of the results would fall between 16 and 29 points. However, a study showed that college basketball games with large point spreads were not symmetric, with 46.2% falling below the spread and 40.7% above the spread.

This result is consistent with games being fixed. However, it is also consistent with the favorite not playing as hard, the favorite pulling the starters out of the game earlier, and other non-sinister explanations. Investigators looked at how the spread in these games changed over time, as a fixed game would attract large bets on the underdog and would therefore cause the point spread to decrease. In fact, it went the other direction, and that may be what caused the asymmetry.

The lesson is to keep looking at the data, testing it in different ways to learn as much as possible.

An important idea to keep in mind is **sampling bias** (or selection bias) in which a bias is caused by circumstances. In looking at the play-by-play for Joe DiMaggio's hitting streak, we find that in situations in which he was 0-3, he got hits in a remarkable 8 out of 9 at bats. Think about it for a second: he usually had 4 at bats during the streak, and we are only looking at games in the streak in which he always got a hit! The remarkable .889 batting average is sampling bias caused by restricting ourselves to games in the hitting streak.

Sampling bias can be subtle. You should be wary any time that you are limiting your data; you may unwittingly introduce a bias.

Here is one more interesting fact from Joe DiMaggio's hitting streak. He struck out only five times in those 56 games. In game 14, the strikeout occurred in his last at bat of the game. In game 16, he followed the strikeout with a double. In games 20, 23, and 24, he followed the strikeouts with home runs! This does *not* look like independent at bats.

Exercises

In these exercises, (T) refers to thinking problems, conceptual problems requiring no calculations. (C) refers to problems requiring significant calculations or calculus. (P) refers to projects; these are ideas for further investigation (hints and resources are at the book's web site).

8.1 For each division in American League baseball in 2014, compute the mean and standard deviations of wins. East: 96,84,83,77,71; Central: 90,89,85,73,70; West: 98,88,87,70,67. Which division was the best? Which division had the most balance?

8.2 If a basketball player's points have mean 30 and standard deviation 4, in how many games of an 82-game season would you expect the player to score (a) more than 34 points; (b) less than 22 points; (c) more than 42 points?

8.3 Two running backs average 4 yards per carry. If player A has standard deviation 2 yards per carry and player B has standard deviation 4 yards per carry, which player is more likely to gain (a) more than 10 yards; (b) less than 0 yards; (c) between 3 and 5 yards?

8.4 Draw a histogram and describe the shape for each. (a) Points scored for the 2014 Dallas Cowboys (17,26,34,38,20,30,31,17,17,31,31,10,41,38,42,44); (b) points scored for the 2014 New England Patriots (20,30,16,14,43,37,27,51,43, 42, 34,21,23,41,17,9). Compare.

8.5 Draw a histogram and describe the shape for each. (a) Wins in the NFL in 2014 (12,9,8,4,11,10,10,7,11,9,3,2,12,9,9,3,12,10,6,4,12,11,7,5,7,7,6,2,12,11,8,6); (b) wins in the NBA in 2014 (49,40,38,18,17,53,50,41,38,32,60,46,37,33,25,51,45,38, 30,16,67,56,39,29,21,56,55,55,50,45). Compare.

8.6 In college football games on a certain date, teams favored by 1, 2, 3, 4, and 5 points won by 12, lost by 4, won by 2, won by 12, and lost by 6, respectively. How often did the favorite win against the spread? Find the differences of spread minus result (e.g., the first difference is $1 - 12 = -11$), and find the mean and standard deviation of the differences.

8.7 Given a standard deviation of 10 points, if a college basketball team is favored to win by 10 points, what percentage of games should it win?

8.8 A team wins 60% of its games. Assuming independence, (a) compute the probability of the team losing 3 out of 4; (b) use the empirical rule to estimate the probability that the team wins 55 or less out of 100.

8.9 A team wins 80% of its games. Assuming independence, (a) compute the probability of the team losing 3 out of 4; (b) use the empirical rule to estimate the probability that the team wins 76 or less out of 100.

8.10 A team wins 35% of its games. (a) Find the probability that it loses its first 21 games. (The 1988 Orioles did this.) (b) Find the probability that this team would lose its first 21 games at least once in 400 tries. (c) How does this compare to estimates of the likelihood of a 56-game hitting streak?

8.11 On June 14, 2015, Andre Iguodala of Golden State (the eventual MVP of the Finals) made 2 out of 11 free throws in game five of the NBA Finals. (a) Using his 70% lifetime free throw percentage, find the probability that he makes 2 or fewer free throws out of 11. (b) Repeat using his season free throw percentage of 60%. (c) How unlikely was this performance?

8.12 A baseball teams wins 56% of its games. (a) Find the expected number of wins in a 162-game season. If it takes 91 wins to make the playoffs and 97 wins to finish in first, use the empirical rule to estimate the probability of (b) making the playoffs; (c) finishing in first.

8.13 In the discussion after Example 8.3, one method of combining win percentages p_1 and p_2 into a win probability was given: $e_1 = \frac{p_1(1-p_2)}{p_1(1-p_2)+p_2(1-p_1)}$. A simpler method is $e_2 = 0.5 + p_1 - p_2$, and a third method (called the "James log-5" method) is $e_3 = \frac{p_1 - p_1p_2}{p_1+p_2-2p_1p_2}$. Show that $e_3 = e_1$. Compare e_1 and e_2 for the following values. (a) $p_1 = .55$, $p_2 = .52$; (b) $p_1 = .60$, $p_2 = .57$; (c) $p_1 = .80$, $p_2 = .77$; (d) $p_1 = .5$, $p_2 = .4$; (e) $p_1 = .55$, $p_2 = .45$.

8.14 Sketch the Lorenz curve, find the Gini index, and compute the entropy for 5-person basketball teams with the following scoring breakdowns. (a) {40, 22, 8, 6, 4}; (b) {20,19,18,17,16}

8.15 Sketch the Lorenz curve, find the Gini index, and compute the entropy for small leagues with the following win breakdowns. (a) {18, 16, 14, 12, 10, 8, 6, 4, 2, 0}; (b) {11, 10, 9, 9, 9, 9, 9, 8, 8, 8}

8.16 For a team with five players and using the computational technique in Example 8.4, find the maximum and minimum Gini index.

8.17 If the luck variance is 40, the skill variance is 400, and the actual variance is 100, find the fraction p of skill.

8.18 (a) If batting averages have mean .270 and standard deviation .030, estimate the probability of a player batting .300. (b) Repeat with a standard deviation of .015. (c) Explain why the second probability is lower.

8.19 Find the means and standard deviations of the top 10 runners in the Olympics women's marathons from 1984 to 2012. Are the winning times improving? Are the mean times improving? Are the standard deviations decreasing? Discuss whether the runners are improving.

8.20 Kobe Bryant made 45% of his shots in his career. In 2011-12 in clutch situations (last 30 seconds, less than 3-point score differential) he made 5 out of 22. Find the probability that with $p = .45$ there would be 5 or fewer successes in 22 tries. Given that everybody in the arena knows that Kobe will take the last shot, is it fair to use his lifetime field goal percentage?

8.21 Use the runs test to test each of the following for independence. (Each has 10 S's and 10 F's.) (a) SFSSSSFFSFSFFSSFFFSF; (b) SSSFFFFSSFSSSSFFFFFS

8.22 Use the runs test to test each of the following for independence. (The numbers of S's and F's are not equal.) (a) SSSSFSSSFSSSSSFSSSSS; (b) SFFSFFSFFFSFFFSFFFFS

8.23 In the 2015 NBA Three-Point Contest, Steph Curry won, recording the following sequence of makes (Y) and misses (N). Use the runs test to test for significance. If he is really a 50% three-point shooter, compute the probability that he would make at least 20 out of 25. NNYYYYYYYNNYYYYYYYYYYYYYN

8.24 In the 2015 baseball Home Run Derby data, batters hit 159 homers in 451 pitches. The sequence of homers formed 206 runs (streaks). Compute the expected number of runs and test for significance. The expected percentages of homer droughts of length 1, 2, 3, and so on are 35, 23, 15, 10, 6, 4, 3, 2, 1, 1. Compare to the actual values of 34, 25, 12, 14, 5, 4, 2, 1, 2, 1. The expected percentages of homer streaks of length 1, 2, 3, and so on are 65, 23, 7, 3, 1, 1. Compare to the actual values of 68, 22, 4, 3, 0, 3. Is there evidence that the hitters were unusually streaky? Winner Joc Pederson hit six homers in a row. What is the probability of him doing that? Why is this small probability not necessarily evidence of hot hands?

8.25 If a player gets hits 35% of the time, find the probability of getting at least one hit in (a) 4 at bats; (b) 5 at bats.

8.26 If a player gets hits 50% of the time, find the probability of getting at least one hit in (a) a game of 4 at bats; (b) 56 straight games with 4 at bats each.

8.27 In Joe DiMaggio's streak, games shortened by rain sometimes limited his at bats. With a batting average of .370, compute the probability of getting at least one hit in five consecutive games with 20 at bats distributed in the following ways. (a) 4, 4, 4, 4, 4; (b) 5, 2, 5, 5, 3

8.28 In the text's analysis of DiMaggio's streak, the 56-game probability of $.84^{56}$ is multiplied by 25 to account for the multiple opportunities for DiMaggio to start a streak. The true probability is $1-(1-(.84)^{56})^{25}$. Explain why this is the correct value, compute it, and compare to the value in the text.

8.29 To see why the issue raised in exercise 8.28 is important, consider taking a basic probability of 0.001 for some event, and asking for the probability that it occurs at least once in 2000 tries. What happens if you just multiply by 2000? Explain why this cannot be correct, and then compute the exact probability.

8.30 Compute BABIP against (assume no sacrifice hits) and DIPS for the given data for Clayton Kershaw and Greg Maddux. How do the BABIP-against values compare to the average BABIP of .300? Is there any evidence that either pitcher was luckier in one season than the other?

	IP	AB	H	HR	BB	HBP	SO
Kershaw 2014	236	851	164	11	54	3	232
Kershaw 2013	237	828	170	16	68	5	229
Maddux 1994	202	734	150	4	34	6	156
Maddux 1993	267	999	228	14	59	6	197

8.31 Compute BABIP for the given data for Tony Gwynn and Jim Thome. How do the BABIPs compare to the average BABIP of .300? Is there any evidence that either hitter was luckier in one season than the other?

	AB	H	HR	SO	SF	SH
Gwynn 1994	419	165	12	19	5	1
Gwynn 1995	535	197	9	15	6	0
Thome 2001	526	153	49	185	3	0
Thome 2002	480	146	52	139	6	0

8.32 (T) For Bernoulli trials with $p \geq .5$, fill in the blank with "larger" or "smaller" and explain: the larger p is, the — the standard deviation is.

8.33 (T) The paradox of skill in Figure 8.10 assumes a fixed limit of human performance, so that skill levels compress. Given the fact that track, swimming, and other records of measurable performance still improve, discuss whether the fixed limit of performance is a reasonable assumption.

8.34 (T) Discuss the ideal balance of skill and luck in a sport.

8.35 (T) In recent years, the standard deviations of batting averages have not decreased. Two explanations could be: (a) baseball players are not close to the limit of human performance; (b) batting average is not selected for, in that managers care more about other skills. Discuss the relative merits of these explanations.

8.36 (T) Discuss the relative values of the great batting averages of Ted Williams in 1941 and Wade Boggs in 1985.

8.37 (T) Explain why, assuming games are independent, the 2-3-2 and 2-2-1-1-1 schedules produce the same winners with the same probabilities. Explain why a 4-3 or 2-1-2-2 schedule would *not* be equivalent.

8.38 (T) According to *Analyzing Wimbledon*, after breaking serve men tennis players hold serve 65.8% of the time, whereas after failing to break they hold serve 64.1% of the time. This looks like a violation of independence, but explain why it could be an example of sampling bias. (Hint: different players have different skill levels.)

8.39 (T) Explain why the probability of each sequence of Hs and Ts in the "Hot Hands" section has the same probability of occurring in coin flipping. Given this, what is the runs test looking for?

8.40 (T) Describe the feeling of being "in the zone." Discuss whether this feeling is the result of a string of successes, or whether the string of successes is the result of the feeling. In other words, which causes which?

8.41 (T) For each of the following, explain what is suspicious about each sequence as the product of fair coin tossing. (a) HHHHHTHHHHHHTTHHHHH; (b) HTHTHTHHTHTTHTHHTHTTH; (c) HHTTTHHHTTHHTTTHHTT

8.42 (T) In the debate about whether Hot Hands exists, describe what is to you the most important piece of evidence.

8.43 (T) While nobody has approached DiMaggio's record of 56 straight games with a hit, Derek Jeter once got hits in 59 out of 61 games, and Johnny Damon got hits in 60 out of 63 games. Do these "near-misses" make DiMaggio's record seem more or less heroic? Discuss.

8.44 (T) Discuss whether the difference between the numbers of hitting streaks of lengths 29 and 30 is convincing evidence of independence being violated.

8.45 (T) Explain why batters have more control over BABIP than do pitchers over BABIP against.

8.46 (T) In 1981, the Clemson University football team went 12-0 and won the national championship. They wore special all-orange uniforms in three important games that year. Discuss the validity of the statement that Clemson was unbeatable in its all-orange uniforms.

8.47 (C) Draw a histogram and describe the shape for each. (a) Points scored and (b) points allowed in games for the 2014-15 San Antonio Spurs; (c) runs scored and (d) runs allowed in games for the 2014 San Francisco Giants.

8.48 Ⓒ Draw a histogram and describe the shape for each. (a) Home runs for American League players in 2014 (players with at least 2); (b) Rebounds per game for NBA players in 2014 (players with at least 2 rebounds per game).

8.49 Ⓒ The pdf for a normally distributed random variable with mean μ and variance σ^2 is $f(x) = \frac{1}{\sigma\sqrt{2\pi}}e^{-(x-\mu)^2/2\sigma^2}$. Take $\mu = 0$ and $\sigma = 1$ and estimate $\int_{-1}^{1} f(x)dx$. Explain how this relates to the empirical rule.

8.50 Ⓒ The shape seen in Figure 8.4 can be described by power law functions of the form $p = cx^{-k}$ and exponential functions of the form $f = ce^{-kx}$. Show that $\ln(p) = \ln(c) - k\ln(x)$ and $\ln(f) = \ln(c) - kx$. Thus, if you have data (x, y) that is from a power law function, a "log-log" plot of $\ln(y)$ versus $\ln(x)$ will be a straight line. By contrast, if you have data (x, y) that is from an exponential function, a "semi-log" plot of $\ln(y)$ versus x will be a straight line. Use log-log and semi-log plots to determine which is power law and which is exponential and determine the underlying equations. (a) {(1,20),(3,2.2),(5,0.8),(7,0.41)}; (b) {(0,20),(2,2.7),(4,0.37),(6,0.05)}

8.51 Ⓒ With a standard deviation of 10 points, how many points better than the spread would you have to be to win 65% of your bets?

8.52 Ⓒ Compute the probabilities in exercises 8(b) and 9(b) exactly, and compare your answers to the values obtained from the empirical rule.

8.53 Ⓒ A team wins 12 out of 20 games. Assuming that games are Bernoulli trials with probability p, write an expression for the probability of winning 12 and losing 8. Think of this as a function of p, compute its derivative and find the value of p that maximizes the probability. This is called a "maximum likelihood estimate" of p.

8.54 Ⓒ Find the Gini index for $f(x) = x^2/100$.

8.55 Ⓒ Show that the maximum entropy for two players occurs when both proportions are 0.5. Assuming that the maximum entropy occurs with equal probabilities, show that the maximum entropy for n players equals $\ln(n)$.

8.56 Ⓒ Find the fractions p of skill for the NBA in 2009-10, 1999-00, 1989-90, and 1979-80 and comment on any trends.

8.57 Ⓒ Find the expected length of the NBA Finals using the 2-3-2 schedule versus the 2-2-1-1-1 schedule assuming (a) all games are independent with the home team winning 65%; (b) all games are independent with the better team (the first one playing at home) winning 75% of its home games and the lesser team winning 55% of its home games.

8.58 Ⓒ In the case of .200-hitter Bob Uecker batting .300 against Steve Carlton, how many at bats would be needed for this result to be statistically significant?

8.59 Ⓒ Pete Rose's lifetime batting average was .303. Assuming independence, calculate the probability that Rose would get (a) at least 13 hits in 30 at bats against Greg Minton; (b) at least 16 hits in 36 at bats against Bob Friend.

8.60 Ⓒ Find the probability that the better team wins a best-of-5 series assuming that: the better team starts at home and wins 75% of its home games, the lesser team wins 55% of its home games, winning the most recent game adds ten percentage points to the chance of winning (e.g., from 75% to 85%), and (a) a 2-2-1 schedule; (b) 1-1-1-1-1 schedule.

8.61 Ⓒ The 1965 New York Giants made a remarkable 4 out of 26 field goals in 1965. Compute the probability of a performance this bad or worse if the Giants had an average kicker (the league made 334 out of 617 field goals in 1965).

8.62 Ⓟ For NFL teams in 1985-2004, compute the Pythagorean wins for each team with exponent 2.37 and the Massey win ratings and identify the best teams. Determine how many times the best team won the Super Bowl. Has this pattern continued since 2004?

8.63 Ⓟ Repeat the Bill James simulation (described following Example 8.3) using the current baseball divisions and schedule. How often does the best team win?

8.64 Ⓟ Find the Gini indices and entropy for all teams from the league of your choice (NFL, NBA, ...). How do these measures of parity correlate to success?

8.65 Ⓟ Simulate numerous seasons of the Luck-NBA to find the variance of wins. Compare your answer to the binomial-based given used in the text.

8.66 Ⓟ Show that there have been more 7-game World Series than a binomial model would predict. Create a model that explains this phenomenon.

Further Reading

Analytic Methods in Sports by Severini gives a thorough development of statistical methods used in sports analysis, with numerous interesting examples.

Nate Silver's *The Signal and the Noise* explores various aspects of separating real patterns from random occurrences in sports, politics, and life.

In "The Relationship Between Concentration of Scoring and Offensive Efficiency in the NBA," Ruiz, Martinez, Lopez-Hernandez, and Castellano use the Gini index to investigate the question of whether star-driven NBA teams are more successful than balanced-scoring teams.

Hot Hand by Reifman covers many of the topics included here. Papers by Tversky and Gilovich and Larkey, Smith, and Kadane are in *Anthology of Statistics in Sports* edited by Albert, Bennett, and Cochran.

56 by Kennedy and *Streak* by Seidel give excellent accounts of Joe DiMaggio's hitting streak.

Bill James' essay "Underestimating the Fog" is a helpful counterpoint to some of the no-hot hands claims.

Triumph and Tragedy in Mudville collects many of Stephen Jay Gould's essays on baseball.

The Information by Gleick introduces many of the ideas in information theory, including entropy. See also *Fortune's Formula* by William Poundstone.

Chapter 9

Sports Strategies

Introduction

FIGURE 9.1: Kevin Kelley?

The opening session at the 2014 Sloan Sports Analytics Conference in Boston featured Bill James (hero of Chapter 7), Hall of Fame NBA coach George Karl, Houston Rockets General Manager Daryl Morey, best-selling author Nate Silver, and Kevin Kelley. In case the name Kevin Kelley does not ring a bell, he is a high school football coach in Arkansas. His teams have won multiple state championships. Still puzzled by his star billing? Kevin Kelley is the Billy Beane of football.

Kelley took a serious look at some of the numbers we will develop in this chapter and dared to ignore convention and implement innovative strategies. His teams, essentially, never punt. They do not try field goals. After touchdowns, they go for two points ... and then try an onside kick. In one important and otherwise close game, Kelley's team led 28-0 before the other team ran an offensive play. This is why Kevin Kelley is a star! Instead of saying that the numbers are nice but nobody is crazy enough to play this way, Kelley puts it on the field and sees what works.

Mathematical analysis sometimes verifies the efficiency of standard sports practices. The challenge for analysts occurs when the mathematics suggests a better way of doing business. When a Billy Beane or Kevin Kelley is willing to put the new ideas into practice, the believers and the skeptics learn what works and what does not. Successes are copied, and failures are re-analyzed.

The numerical logic behind the strategies in sports is the topic of this chapter. We will attempt to answer the following questions. Is Kevin Kelley crazy? What about Bill Belichick? Is bunting a good strategy in baseball? What is the best strategy for penalty kicks? How much does a team's batting order matter? What do "leverage" and "game control" mean? What does John Nash have to do with sports? How important is it for teams to have star players? At what age do baseball players hit the most home runs?

Don't Punt, John!

Your team faces fourth down with two yards to go at midfield. Which option do you choose, go for it or punt? You already know Kevin Kelley's answer, but let's see in what way it is correct.

We start with some numbers from pro-football-reference.com. In 2014, NFL teams went for it on fourth down and 2 or 3 on 83 occasions, and were successful 53% of the time. The average gain was 6 yards. So, let's say your team is successful 53% of the time, and you either get a first down at your opponent's 40 or they take over the ball at the 50. The alternative is to punt. In 2014, the 107 punts from the 50 averaged 40 yards with a return of 6 (no information on how many touchbacks). Let's say the outcome of a punt is for the opponent to take over on its 16.

How do we evaluate which is the better strategy? One way is to compute an expected value of the score. The relevant research, amazingly enough, dates back to Virgil Carter, a quarterback for the Cincinnati Bengals who published a mathematics paper in the journal *Operations Research* in 1971. The idea is to track every drive for several seasons and note the next score. For example, a team starts a drive on its own 20 and scores 7 points; another team starts a drive on its own 20, does not score, but its opponent scores 3 points on its next possession. At this point, 2 drives from the 20 have netted +4 points (7-3) for an average of +2 points. Doing this for every yard line and several years gives a good idea of how many points a position on the field is worth. It turns out that the data is very close to forming a straight line connecting 6 points when 1 yard from the goal line (ask the 2015 Super Bowl teams why it is not 7) to -1 point (the other team is more likely to score next) at 100 yards from the goal line. The function $p(x) = 6 - .07x$ gives an approximation of the expected points when x yards from the goal line.

Example 9.1 Compute the expected values for going for it and punting from the 50 on 4th down and 2.
Solution Recall that expected value equals the sum of probability times value for all possible outcomes. Going for it, we have two possible outcomes. Making a first down has probability 53% and value $p(40) = 3.2$, the expected number of points from a first down at the opponent's 40. Getting stopped has a probability of 47% and value $-p(50) = -2.5$, the negative of the opponent's expected points from a first down at the 50. After a punt, the expected value is $-p(84) = -.12$. The expected values are
Going For It: $.53 * 3.2 - .47 * 2.5 = 0.521$
Punting: -0.12
The expected value is larger (and positive) for the strategy of going for it. (As Kevin Kelley could have told us.)

A coach would be considered a reckless gambler going for it at the 50. What about going for it inside your own territory?

Example 9.2 Compute the expected values for going for it and punting from your own 30 on 4th down and 2.
Solution We will use the 53% success rate from Example 9.1, and assume that the options are first down at the 40 or the opponent's ball at your 30. From pro-football-reference.com, punts from the 30 in 2014 averaged 43 yards with a 9-yard return, giving the opponent the ball at its 36.
Going For It: $.53 * p(60) - .47 * p(30) = -0.88$
Punting: $-p(64) = -1.52$
The expected value for going for it is now negative, but it is less negative than the punting value. Therefore, you should go for it!

This is starting to look crazy. Let's phrase that differently: either pro and college football coaches are choosing ineffective strategies, or there is something wrong with these calculations. The importance of Kevin Kelley is that he has shown us that, at his high school level, the calculations are largely correct. It is important to note that the lower quality of punting in high school makes Kelley's refusal to punt even sounder.

Of course, the calculations grossly oversimplify the decisions that coaches make. Offensive and defensive match-ups vary, the score and time remaining are important considerations, and so on. If teams start going for it routinely, the success rates will change. Nevertheless, Examples 9.1 and 9.2 raise interesting questions.

In a side note, Virgil Carter had a starring role in a strategy situation that changed professional football (as told in the book *Newton's Football*). Carter was backup to Greg Cook, who was having a sensational rookie season running Bill Walsh's offense in Cincinnati. When Cook was injured, Carter took over but did not have the arm strength to make the necessary throws. In desperation, Walsh shifted the offense to a Carter-friendly short passing game that was the genesis of the West Coast offense that Walsh perfected in San Francisco.

Bill Belichick's Gambles

If you are a long-time New England Patriots or Indianapolis Colts fan, you recognized the scenario in Example 9.2. In 2009, with a 34-28 lead and only two minutes remaining, Patriots Coach Bill Belichick went for it on fourth down and two to go. The Patriots did not get the first down and lost the game 35-34 on a late touchdown pass. Example 9.2 indicates that the decision was not dumb, especially given the Patriots' 75% rate of converting fourth downs.

In the 2012 Super Bowl, Belichick's Patriots led 17-15, but the New York Giants had the ball at the New England 6 and the clock was approaching one minute remaining. Belichick ordered his defense to let the Giants score a touchdown, gambling that his offense could score a touchdown in one minute to retake the lead. They did not, and the Giants won. The key issue here is not expected points, but the probability of winning the game. A field goal would have won the game for the Giants and, other than a highly unlikely fumble or missed field goal, there was no scenario in which the Patriots could win the game if the Giants had the ball.

As with expected points, win probabilities can be computed if enough data is available. Unlike expected points, win probabilities require several variables. With expected points, it makes some sense to say that teams average 2.5 points when starting 50 yards from the goal line. The score and amount of time left are not necessarily important (an obvious exception being if only two seconds are left). However, it makes no sense to ask what the probability of winning is when you start 50 yards away; you have to know the current score and time remaining, also. This makes win probabilities harder to approximate; there is no simple $p(x)$ formula. As Advanced NFL Stats (later renamed Advanced Football Analytics) computed it, the Giants had a 94 percent chance of winning before the play. After the Patriots let them score, the probability of the Giants winning dropped to 85 percent. The Patriots still had little chance to win, but the odds were better.

Interesting plays seem to follow Belichick, because in the 2015 Super Bowl there was a near replay of the 2012 game. The Patriots led the Seattle Seahawks 28-24 with just over one minute remaining, but Seattle had the ball at the New England 1. Instead of letting Seattle score or calling a timeout to preserve time, Belichick allowed the clock to run. In a stunning finish, the Patriots won the game when Malcolm Butler intercepted a pass on the next play. In this case, Seattle had an 88 percent chance of winning before the play. This would have dropped to 77 percent if the Patriots had called timeout and Seattle scored a touchdown. (It would have been less than the Giants' 85 percent because in this case Seattle would have led by three points, so that a Patriots field goal could have tied the game.)

The evidence is that Belichick made the right call in the first two cases, and lost, and made the wrong call in the last case, but won. Probabilities are, after all, only probabilities.

What about Seattle's decision to try a pass? Here are some numbers from pro-football-reference.com. On plays with two or less yards to go for a touchdown, in 2014 teams ran the ball 333 times and threw 208 times. Runs produced touchdowns 54.1% of the time, and turnovers 1.5% of the time. Passes produced touchdowns 52.4% of the time, and only one turnover (a fumble). Thus, calling for a pass was neither unusual nor dumb, especially given the amount of time remaining. The outcome was unfortunate for Seattle.

The Value of a Play

The concepts of expected points and win probability give us different ways of evaluating plays and players. For example, a team completes a 70-yard pass from its own 20 to its opponent's 10. Instead of calling this 70 yards, we can say how many points it is worth. Before the play, the offense had expected points of $p(80) = 0.4$ and after the play expected points increase to $p(10) = 5.3$. The play is worth $5.3 - 0.4 = 4.9$ points.

Three plays later, the team's running back scores a touchdown from the 1-yard-line. He gets credit for the touchdown, but did not actually add much to the expected score. Using the basic expected points formula, he added 1 point (from an expected 6 to an actual 7).

Win probabilities could tell a different story. If the 70-yard pass play was at the end of a 35-0 blowout, then the win probability would not change; the probability is 100% at 35-0, and doesn't get any higher at 42-0. If it were at the end of a game that was tied, it would increase the probability of winning from about 50% to about 90%. The WPA (win probability added) points would be 40, reflecting a play with a large influence on the outcome of a game.

Notice that the play would receive a large number of WPA points even if the team lost the game due to a subsequent fumble or missed field goal.

If the expected number of points follow a straight line, a counterintuitive conclusion follows, as shown in Carroll and Palmer's book *The Hidden Game of Football.*

Example 9.3 Use the expected points formula $p(x) = 6 - .07x$ to compute the cost of losing a fumble at (a) your own 10-yard-line; (b) your opponent's 10-yard-line; (c) x yards from the goal line.

Solution (a) At $x = 90$ yards from the goal line, expected points equal $p(90) = -.3$. The opponent gets the ball 10 yards away with expected points $p(10) = 5.3$ which is an expected -5.3 points from your perspective. The difference is $-5.3 - -.3 = -5.0$ points. (b) At $x = 10$ yards away, you drop from an expected 5.3 points to an expected .3 points, a difference of $.3 - 5.3 = -5.0$ again. (c) For any x, you go from $p(x)$ to $-p(100 - x)$, with a difference of $-p(100-x) - p(x) = -[6 - .07(100-x)] - [6 - .07x] = -6 + 7 - .07x - 6 + .07x = -5$, exactly as we computed in parts (a) and (b).

Is it true that a fumble costs 5 points no matter where on the field it occurs? It certainly doesn't feel that way to fans. Perhaps the emotional impact of a fumble near the goal line changes the expected points formula. Otherwise, the formula forces us to conclude that a fumble is a fumble. As you will show in the exercises, any linear function for $p(x)$ will force the same qualitative conclusion.

Markov Chain Models

In this section, we introduce a mathematical technique that proves valuable for analyzing some sports. Baseball is an orderly sport that lends itself to mathematical analysis. There are a small number of states, or situations, that describe all possibilities in an inning. There are eight base situations (no runners, runner on first, runner on second, runners on first and second, and so on) which can be paired with the 3 out possibilities to make 24 states. Adding the "inning over" possibility makes 25 states. An example of a state is "1 out, runners on second and third."

A **Markov chain** is a set of states and transition probabilities that describe movement between states. There is, as usual, an assumption of independence, that past results do not affect the probabilities. By studying Markov chains in general, mathematicians have discovered important formulas to simplify calculations. Here, we will work with a small example to see how this might work.

We invent a simple version of cricket/baseball, let's call it OneBase, in which each batter either gets a single, a homer, or an out. A single puts the batter on base; if there was a runner on base, he scores. A homer counts as one run (two if a runner is on base). An out is like a strikeout; a runner does not advance. An inning consists of two outs, and there is only one base other than home. There are five states: (1) 0 outs, 0 on base; (2) 0 outs, 1 on base; (3) 1 out, 0 on base; (4) 1 out, 1 on base; (5) 2 outs, inning over. This last state is called an **absorbing state** since once the second out is recorded nothing else happens; the process has been absorbed into the fifth state forever.

Assume that a homer occurs with probability .1, a hit with probability .2, and an out with probability .7. The first task is to compute the **transition probabilities**, the probabilities of changing from one state to another. Start with T_{11}, the probability of starting in state 1 (0 outs, 0 on base) and staying in state 1. The only way this can happen is if the batter hits a homer, which happens with probability .1. So $T_{11} = .1$. Next, T_{12} is the probability of starting in state 1 and ending in state 2; this happens if the batter hits a single, with probability $T_{12} = .2$. The transition from state 1 to state 3 occurs if the batter makes an out, so $T_{13} = .7$. It is not possible to move directly (in one batter) from state 1 to state 4 or state 5, so $T_{14} = T_{15} = 0$.

Even with our reduced set of states, this can get tedious. We store all of our probabilities in a matrix. The five numbers we just computed go in the first row of the matrix. So T_{ij} will refer to the number in the i-th row and j-th column. Note that this means that if you add the numbers in any row, you will get a sum of 1. We have

From, To	0 out, 0 on	0 out, 1 on	1 out, 0 on	1 out, 1 on	over
0 out, 0 on	.1	.2	.7	0	0
0 out, 1 on	.1	.2	0	.7	0
1 out, 0 on	0	0	.1	.2	.7
1 out, 1 on	0	0	.1	.2	.7
over	0	0	0	0	1

which without the labels can be summarized in the matrix T given by

$$T = \begin{bmatrix} .1 & .2 & .7 & 0 & 0 \\ .1 & .2 & 0 & .7 & 0 \\ 0 & 0 & .1 & .2 & .7 \\ 0 & 0 & .1 & .2 & .7 \\ 0 & 0 & 0 & 0 & 1 \end{bmatrix}$$

which is called the **transition matrix**. Take a second to follow the logic in each line of the matrix. The state changes with each batter, either through a homer, a single, or an out. Thus, you see the three associated probabilities in each line except the last. For example, the fourth row describes what happens starting with 1 out and 1 on. A homer turns the state into 1 out and 0 on (two runs score), a single leaves the state at 1 out and 1 on (one run scores), and an out is the second and last out of the inning. The last row simply says that when the inning is finished, it remains finished.

One quantity of interest here is runs scored. To see how to get runs involved, consider the second row in which the initial state is 0 out and 1 on. A homer (probability .1) scores 2 runs, and a single (probability .2) scores 1 run. The expected number of runs for one batter from this state is 2x.1 + 1x.2 = .4. The only way to score from states 1 and 3 is a homer, so the expected number of runs for one batter from states 1 and 3 is .1. The expected runs scored for one batter can be summarized in the vector

$$R = [r_1, r_2, r_3, r_4] = [.1, .4, .1, .4, 0]$$

This was not hard to compute, but the quantity that we really want is the expected number of runs scored from a given state for the *entire* inning. This calculation will look like the matrix calculations in Chapter 5 for the Massey ratings.

Start by assigning names to the expected number of runs from each state: we can use a, b, c, d, and e. Think through the process of scoring runs from the first state, 0 out and 0 on. Runs can be scored from the first batter (what we computed above) or by scoring runs starting with the second batter. Then

a, the total expected runs scored, equals the initial value of .1 plus subsequent runs scored from any of the other states. From state 1, the transition to state 1 has probability T_{11}, and from state 1 we expect to score a runs; the transition to state 2 has probability T_{12}, and from state 2 we expect to score b runs; the transition to state 3 has probability T_{13}, and from state 3 we expect to score c runs; the transition to state 4 has probability T_{14}, and from state 4 we expect to score d runs. All in all, we have

$$
\begin{aligned}
a &= r_1 + T_{11}a + T_{12}b + T_{13}c + T_{14}d \\
b &= r_2 + T_{21}a + T_{22}b + T_{23}c + T_{24}d \\
c &= r_3 + T_{31}a + T_{32}b + T_{33}c + T_{34}d \\
d &= r_4 + T_{41}a + T_{42}b + T_{43}c + T_{44}d
\end{aligned}
$$

where the equations for b, c, and d are derived in the same way.

From here, you can get a computer or calculator to solve the equations for you. (In linear algebra terms, the solution is $(I - S)^{-1}R$ where I is the 4x4 identity matrix and S is the upper 4x4 submatrix of T.) The solution is $a = .52$, $b = 1.03$, $c = .23$, $d = .53$.
The best scoring state is the second state with 0 out and 1 on, and the worst scoring state is the third state with 1 out and 0 on. This is logical.

The same process can be done for baseball. There are various issues for getting the best model possible. Instead of adding one state for "inning over" you can allow runs to score on the last out by making four extra states corresponding to the last out made with 0 runs scoring, the last out made with 1 run scoring, and so on. Further tweaks can account for stolen bases, balks, and other events that do not result in a change of batter. The probabilities in the transition matrix can be estimated simply (find the proportion of at bats resulting in a double and assume that all runners advance two bases) or more fully (e.g., the probability of a double could depend on the current state).

Several of the statistics that can be computed from Markov chain models of baseball agree closely with league statistics. This serves to validate the model, which might be dubious given an assumption of independent at bats. With the model in place, experiments can be run to test strategies. One interesting result that has been explored by several researchers is the effect of batting order in baseball. The bottom line is that it does not seem to matter much, with the difference between best order and worst order being 30 to 40 runs (3 or 4 wins). The general principle is that you want the best batters to bat near the top of the order, so that they get the most at bats. As Bill James has said, the main difference in most reasonable batting orders is in which innings the runs will be scored.

The Expected Runs Matrix

Once you have the average number of runs scored from each state, you can analyze in-game strategies. From an expected runs matrix found online, you can answer questions like the following, where the data comes from Tom Tango's website.

Example 9.4 Find the expected run values for a runner on first with 0 outs and a runner on second with 1 out. Discuss whether a sacrifice bunt is a good play.
Solution At tangotiger.net, the expected runs scored with 0 outs and a runner on first from 1993-2010 is 0.941, while the expected runs scored with 1 out and a runner on second is 0.721. The team is better off in the first situation, so a sacrifice bunt that puts the team in the second situation is an ineffective play.

There are numerous problems with the analysis in Example 9.4. The tables give an average/expected number of runs over multiple seasons. If the batter is below average (for example, a pitcher), the bunt may be a good play. This is explored in the exercises. A different objection is that the goal may not be to score the most runs. If it is a 3-3 game in the bottom of the ninth, the team only needs one run to win. Instead of looking at the expected number of runs, the probability of scoring one run is more relevant. Tango's website includes that information as well. Teams score from a runner on first, 0 out state 44.1% of the time, compared to 41.8% of the time from the runner on second, 1 out state. The bunt still looks like an inferior play, in general.

Win Probability and Leverage

The concept of win probability is straightforward and has been referred to several times. Instead of looking at how many points or runs are scored from a particular state, we look at the probability of winning from a particular state. As previously discussed, this means we need to account for score and time remaining. In baseball terms, this means looking at score and inning.

Example 9.5 Find the win probability for a runner on first with 0 outs and a runner on second with 1 out, in the bottom of the ninth with the score tied. Discuss whether a sacrifice bunt is a good play.
Solution At tangotiger.net, the win expectancy (probability) for innings 7-9 and run differentials of 0 or 1 are given. With 0 outs and a runner on first

in the bottom of the ninth with the score tied, the probability of the home team winning is about 0.715. The win probability with 1 out and a runner on second is 0.703. It is close, but the team is better off in the first situation, so a sacrifice bunt that puts the team in the second situation is an ineffective play.

Win probabilities allow us to measure how critical a play is, or how much **leverage** a play has on the outcome. A simple form of the Leverage Index can be computed using the situation in Example 9.5 with the batter swinging. There are multiple possible outcomes. Let's say that this pitcher-batter matchup produces home runs 10% of the time, singles 20% of the time, and simple outs 70% of the time (no double plays, or runners advancing). The team starts with a win probability of .715, which is raised to 1 (they win!) on a home run. This is a change in win probability of .285. A single raises the win probability to .816, an increase of .101. An out lowers the win probability to .637, a decrease of .078. The average change (ignoring whether the change is positive or negative) is $.1(.285)+.2(.101)+.7(.078) = .1033$. Tango estimates the average change in win probability during the course of a game to be .0347, so our situation is $\frac{.1033}{.0347} \approx 3$ times as important as a normal play. The leverage index of 3 quantifies the significance of a situation, and is useful in quantifying clutch play.

Game Control and the Story Stat

On May 19, 2015, the Minnesota Twins beat the Pittsburgh Pirates 8-5 and the Washington Nationals beat the New York Yankees 8-6. Based on the scores, the games seem to have been similar. In fact, the Nationals rallied from a 6-2 deficit and won on a walk-off home run. The Twins took an early 8-1 lead and cruised to an easy win, giving up meaningless single runs in the fourth through seventh innings. We can quantify the closeness of the games using win probabilities.

At baseball-reference.com, the play-by-play game score includes win probabilities after each at bat. Danny Santana opened the game for Minnesota by grounding out; this lowered the Twins' chances from 50% to 48%. Brian Dozier homered to give the Twins a 1-0 lead, raising the win probability to 59%. In the second inning, Joe Mauer hit a bases loaded double to make the score 5-0, with a win probability of 90%. The win probability never dropped below 83% the rest of the game. The average of these win probabilities gives an idea of the closeness of the game. For the Twins, add 48 and 59 and so on and divide by the total number of at bats. The result is 89.1, showing that for most of the game the Twins were in a commanding position. By contrast, the average for the Nationals-Yankees game, a back and forth affair, is 52.3, showing that the Nationals were not in a strong position for much of the game.

The graphs of the win probabilities for the two games gives a visual of the difference in the games, belying the similar final scores. *Baseball Prospectus* calls this type of graphic the "Story Stat" because of the immediate visual evidence of the ebbs and flows of the game (or, in the case of the Twins-Pirates game, the lack thereof). The sudden impact of the Nationals' walk-off home run is immediately apparent.

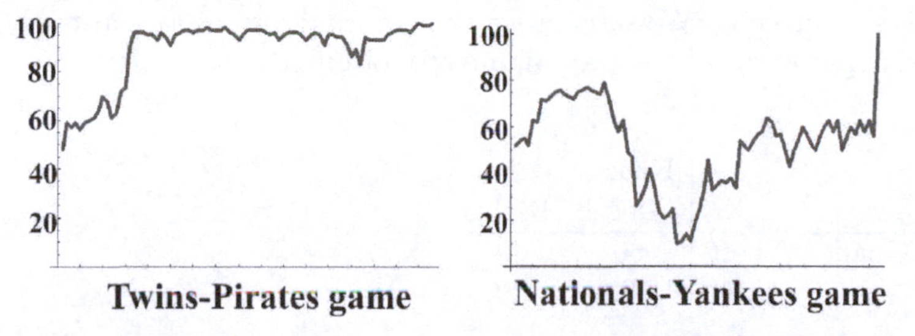

FIGURE 9.2: Win Probabilities for Twins and Nationals

The selection committee for the 2015 college football playoff received some unwanted publicity for using average win probabilities, which they called "game control," as a measure of teams' strengths. Florida State, which had fallen behind by double-digit points in several games before winning, was particularly unhappy at having this aspect of its play quantified.

Game Theory

Similarities between baseball pitching and tennis serving include launching projectiles at high speeds at opponents who have less than a half-second to judge the trajectory of the ball and make contact with it. Pitchers and servers vary the speed, spin, and location of their missiles to keep their opponents off-balance and guessing. The mathematical analysis of this type of cat-and-mouse game is called game theory.

Game theory gained some cultural popularity from the movie *A Beautiful Mind* about one of game theory's pioneers, John Nash. The name "game theory" does not send people the right message. James Case titled his book on game theory *Competition*, and competition theory is a much more accurate name. Game theory looks at "zero-sum games" in which whatever happens positively for one competitor happens negatively to the other, and "non-zero-sum games" in which cooperation can be a positive. The most direct applications to sports are the zero-sum games.

Penalty kicks in soccer are well modeled by game theory. The kicker has options on where to kick the ball. The goalie does not have time to see where the ball is going and react, and so must choose a direction to move. To show one possibility in game theory, consider a young player who kicks right-footed. His or her kicks are stronger and more accurate when aimed to the left. If the goalie guesses correctly that the ball is going to the left, the kick is successful 60% of the time. If the goalie guesses incorrectly, the kick to the left is successful 100% of the time. However, if the kicker goes to the right, the kick is successful only 20% of the time when the goalie guesses correctly and 40% when the goalie guesses incorrectly. We summarize this information in, as you might guess, a matrix that is called the **payoff matrix** of outcomes.

		Kicker	
		Left	Right
Goalie	Left	60	40
	Right	100	20

For this payoff matrix, the label "Left" for the goalie means that the goalie is guessing that the kicker will kick the ball to the left (which is the goalie's right).

Example 9.6 Determine the best strategies for the kicker and goalie with the payoff matrix $\begin{bmatrix} 60 & 40 \\ 100 & 20 \end{bmatrix}$

Solution Mathematically, we analyze this game in the following way. From the kicker's perspective, whether the goalie guesses left or right, the percentage is higher if the ball is kicked to the left. The left column dominates the right column, in that each entry is larger. Therefore, the proper strategy is a "pure strategy" of always kicking to the left. From the goalie's perspective, the worst that can happen guessing left is for the kicker to go left and score 60% of the time. The worst that can happen guessing right is for the kicker to go left and score 100% of the time. The best of the worst (this is called the "minimax" since we are choosing the smaller of the two larger numbers) is to guess left, which is the goalie's pure strategy.

In sports terms, we come to the same conclusion. The kicker gets a better result going to the left, no matter what the goalie does, so the kicker should always go left. The goalie, giving the kicker credit for making the right choice, knows that the kicker is going left and so must guess left to minimize the damage.

Let's make Example 9.6 more realistic by changing the payoff matrix to

		Kicker	
		Left	Right
Goalie	Left	64	89
	Right	94	44

The kicker is still stronger going to the left, being successful 64% of the time when the goalie guesses correctly. When the kicker goes right and the goalie guesses correctly, the success rate is only 44%. However, if the goalie guesses left, the kicker is better off fooling the goalie by going to the right. It is now a guessing game, although an interesting one because the kicker would prefer going to the left. But the goalie knows that and so will tend to guess left, which would make the kicker go to the right, ..., so that we have quite a logical web to untangle.

The first main point to make here is that the players do not have a pure strategy to use. There is not a dominant row or column. Further, either player can take advantage of knowledge of the other player's strategy. That's our logical tangle: if the goalie knows the kicker is going left, the goalie will go left, but if the kicker knows the goalie is going left, the kicker will switch to going right, and so on.

This leads to the second main point: the optimal strategy is a "mixed strategy" in which each player uses each of the available strategies a fraction of the time. The challenge is to figure out the best fractions: should the kicker go left 50% of the time? 60%? How often should the goalie guess left?

The basic principle for determining the optimal percentages is that a given player does not want the other player to be able to gain an advantage by knowing the strategy. Both players, in theory, should be able to announce their percentages and know that the other player cannot use the information to improve the odds. Example 9.7 shows how this can be done.

Example 9.7 Determine the best strategies for the kicker and goalie with the payoff matrix $\begin{bmatrix} 64 & 89 \\ 94 & 44 \end{bmatrix}$.

Solution Start by giving names to the variables we want to find. Let k be the fraction of time the kicker goes left and g the fraction of time the goalie guesses left. Then the kicker goes right $1-k$ percent of the time and the goalie guesses right $1-g$ percent of the time. (If that is not clear, think of an example. If the kicker goes left 40% of the time, then the kicker goes right 60% of the time, and we have fractions $k = .4$ and $.6 = 1-k$.) Replacing the strategy names with the probability names will give us a handy reference.

		Kicker	
		k	$1-k$
Goalie	g	64	89
	$1-g$	94	44

First, compute the expected outcome if the kicker goes left. The kicker going left is the first column of the payoff matrix, so look at the two numbers in the first column and the corresponding probabilities to their left. The outcome 64 occurs with probability g and the outcome 94 occurs with probability $1-g$. The expected value is $64g + 94(1-g)$. Next, compute the expected outcome if the kicker goes right. This is the second column of the payoff matrix, so match the numbers in the second column with the probabilities to the left. The outcome 89 occurs with probability g and the outcome 44 occurs with probability $1-g$. The expected value is $89g+44(1-g)$. The basic principle is that these need to be the same, so we set them equal to each other and solve. From $64g+94(1-g) = 89g+44(1-g)$ we get $64g+94-94g = 89g+44-44g$ or $50 = 75g$ and hence $g = \frac{50}{75} = \frac{2}{3}$. The goalie should guess left two-thirds of the time and right one-third of the time.

Switch roles and compute the expected outcome if the goalie guesses left. This is the first row of the payoff matrix, so match the numbers in the first row with the corresponding probabilities above them. The outcome 64 occurs with probability k and the outcome 89 occurs with probability $1-k$. The expected value is $64k + 89(1-k)$. If the goalie guesses right, we use numbers from the second row of the payoff matrix and the probabilities above. The outcome 94 occurs with probability k and the outcome 44 occurs with probability $1-k$. The expected value is $94k+44(1-k)$. The basic principle is that these need to be the same, so we set them equal to each other and solve. From $64k + 89(1-k) = 94k + 44(1-k)$ we get $64k + 89 - 89k = 94k + 44 - 44k$ or $45 = 75k$ and hence $k = \frac{45}{75} = \frac{3}{5}$. The kicker should go left three-fifths (60%) of the time and right two-fifths (40%) of the time.

If both kicker and goalie follow the optimal strategies, the fraction of successful penalty kicks can be found by substituting in g or k into one of the original expected values. For example, setting $k = .6$ in the formula $64k+89(1-k)$ gives an expected value of .74: the kicker converts 74% of the kicks.

Example 9.7 is still a major simplification of the penalty kick situation. A study reported on by Oliver and Wilson divides the net into nine regions with left-middle-right horizontally matched with low-middle-high vertically. The results are interesting. Going high is high risk/reward, with several shots going too high but with no goalie saves on high shots in the study. Over 70% of the saves occurred on low kicks. A study reported at scienceofsocceronline.com indicates that the goalie dives left or right on 94% of shots. Kickers going down the middle were successful 87% of the time, compared to 83% overall success.

A third study published by Chiappori, Levitt, and Groseclose looked only at horizontal placement, divided into left-middle-right. They noted that save rates were nearly equal when the goalie guessed left or middle or right. The actual percentages are not close to optimal, but having equal save percentages is a property of the optimal solution and indicates how athletes often find optimal solutions. If the save percentage was low in one area, goalies would go in that direction less often, making their left/middle/right breakdown closer to the ideal.

Upsetting the Game Theory

The basic philosophy of game theory is very conservative. Strategies minimize the worst outcome that can occur, and guard against knowledge of the strategy being exploitable. This does not mean that game theory recommends conservative strategies.

In football, there has long been a theory that to win, a team must establish the run. Indeed, in 2014 the correlation between rushing yards and wins in the NFL is a solid 0.4. Recall that correlation does *not* imply causation; in other words, correlation shows that winning teams tend to have high rushing totals, but it does not mean that a team will win more if they run more. In fact, the 2014 correlation between rushing attempts and wins is a puny 0.05.

The modern football offense has a mixture of runs and passes, and it is sometimes said that a team must establish the run to be able to pass. This is partially in line with game theory: mixed strategies should be used randomly so that the opponent has no idea what is to happen next. The full game theory lesson is that you must establish both run and pass to be able to run and pass. The opponent needs to know that you might run up the middle or throw a deep pass.

There are times when the measurement criterion is not expected score but probability of winning. There may be some games where losing by 6 is much better than losing by 30, but there may also be games where all you care about is winning. Then you might want to move away from the game theory strategies and try something different. The conservative long-term strategies of game theory are great for the better team, but how can the underdog team pull off an upset?

The concept of a high-variance strategy is important. To put this in basketball terms, suppose your team is a 10-point underdog, with mean scores of 60 for and 70 against. If each score has a standard deviation of 3, then the ranges of likely scores are 54-66 and 64-76, with almost no overlap for an upset to occur. If, instead, each score has a standard deviation of 6, then the ranges of likely scores are 48-72 and 58-82. There is now a much larger range of overlap and a better chance of an upset.

Peter Keating and Jordan Brenner of *ESPN The Magazine* have a formula for identifying possible upsets that relies heavily on high-variance characteristics such as pressing defenses, reliance on three-point shooting, and other risky strategies. Interestingly, in recent years several teams that would have been likely "giant killers" in the NCAA Tournament were themselves upset in their conference tournaments and did not reach the NCAA Tournament. High-variance strategies cut both ways.

Getting and Giving Two

A basketball team gets the ball late in a quarter, with about 15 seconds more on the game clock than on the shot clock. The television announcers immediately start talking about going "two-for-one." This strategy calls for a quick shot, so that there is time to get the ball back a second time at the end of the quarter. Without the quick shot, the other team gets the last shot of the quarter. For example, in the NBA with its 24-second shot clock, taking a quick shot with 30 seconds left in the quarter means that you will get the ball back with at least 6 seconds left (assuming no turnovers or offensive rebounds). If you don't shoot until 20 seconds are left, you might not get the ball back.

A quick calculation makes the two-for-one look like the right strategy. Suppose that you average 1 point per possession. Taking two quick shots might lower that efficiency by 20% to 0.8 points per possession. Nevertheless, you can expect 1.6 points on your two quick possessions as opposed to 1 point for one long possession. Reality makes this far more complicated. If you take your time and shoot with 20 seconds left, the other team will often take a quick shot and leave you 4 or 5 seconds for the last shot. Offensive rebounds and turnovers occur with some regularity, as well.

In *The House Advantage*, Jeffery Ma cuts through this theoretical knot with data. For a given time remaining in the quarter, he computed the average score for the rest of the quarter. With 45 to 120 seconds left, the offense has a 0.5-point advantage. With 3 to 20 seconds left, the offense has a full point advantage. The smallest advantage is 0.25 points with 33 seconds. This is an ideal time to shoot, when a change of possession is least costly.

Another end-of-game situation that creates debates is when a team is ahead by 3 points with time running out. Should they let the other team attempt a three-point shot to tie the game, or foul them and make them shoot two free throws? Again, a quick calculation gives a decisive answer that is not necessarily correct. Suppose the other team makes 35% of its three-point attempts. The game gets tied 35% of the time if they are allowed to shoot the three. If fouled, the other team needs to make a free throw (let's say 80%), intentionally miss the next free throw but get the rebound (maybe 60%) and then make a shot to tie (maybe 50%). Multiplying the probabilities, the team has a probability of $.8x.6x.5 = .24$ of accomplishing the full chain of events, much smaller than the 35% chance. Therefore, foul them. However, if you foul them with 6 seconds left, the other team has time to foul you right back and then get the ball back with 5 or so seconds left, plenty of time to get a good shot (and, if you missed one of your free throws, they don't have to shoot a three this time). Plus, if they shoot and make a three-pointer with 6 seconds left, then you have a chance to win the game with a last shot. The final decision is not clear.

The Physical Challenge

Not all issues of strategies in sports are easily quantified and analyzed. A couple of interesting situations are discussed here.

In the 2008 Olympics, Michael Phelps came from behind in the 100 m butterfly to win by .01 second (the limit of the timing at the Olympics) over Milorad Cavic. Phelps took what most coaches consider an ill-advised last stroke before hitting the wall, while Cavic coasted into the wall. Phelps's stroke was perfectly timed and his fingertips hit the wall just before Cavic's.

A similar situation occurs frequently in baseball when a runner tries to beat a throw to first base by diving to the bag. Is it better to dive than to run all the way through the bag? It is mostly an issue of timing. Experiments tend to find diving slower, but it depends on whether the runner dives onto the bag (fast, if dangerous for jamming a finger) or hits the ground first and slides into the bag (slow). As for running, if the runner's natural rhythm brings a foot down onto the bag, that is faster. If the runner must shuffle his feet to hit the bag, that is much slower.

Reaction times were discussed briefly in previous chapters. In a race, is it worthwhile trying to anticipate the start of the race to get a slight edge on the competition? In swimming and track, the penalty for a false start (starting early) is typically disqualification. On the track, pressure plates in the starting blocks can determine when a sprinter begins to run, and anyone who starts less than 0.1 s after the gun fires is deemed to have false-started. This is based on extensive research into optimal reaction times to auditory signals, which indicates that a reaction time of 0.12 s is near the limit of human abilities.

Reaction time should be positively correlated with overall time: the sooner you start, the sooner you finish. However, data published by the IAAF consistently shows little or no correlation between reaction time and overall time. The explanation could be that you can become an elite sprinter by having a fast start or by having a fast finish, and the long legs that can help you run fast will slow down your start.

Reaction time is different in drag racing. The timing mechanism is precise and predictable, a series of lights that flash at regular intervals. Drivers can train themselves to start at exactly the right time. An episode of *Sports Science* clocks Hillary Will with a reaction time of 0.001 s. Clearly, she is not reacting but instead is anticipating. Precision timing is obviously a skill that drag racers need.

Personnel Decisions: Aging

Some of the most important analytics work being done supports personnel decisions. Much of this is proprietary, complicated, or not that interesting to the common fan. A brief discussion follows on three aspects of this area of analytics.

The concept of a peak age is simple enough to understand. Young athletes have not fully honed their craft, while old players are fighting against physical decay. Somewhere in the middle is the age at which the best performances occur. The question is how to determine this peak age. But, let's be more precise, more mathematical. What exactly are we measuring? The peak age for batting average could be different for the peak age for home runs, and so on. You could look at a specific skill or some measure like WAR of overall performance.

Having decided on a statistic to track (I'll use home runs as an example), we next decide what data to collect. We could pick a year like 2014 and find how many home runs were hit by 20-year-olds, then 21-year-olds, and so on. Figure 9.3 shows the result. We could conclude from this that the peak age for home runs is 28. But, there might be more home runs hit by 28-year-olds because there are more 28-year-olds playing. The large drop in home runs at age 29 is more likely due to fewer 29-year-olds in the league than to 29 being a jinxed age. So, we should control for that by computing home runs per 500 at bats in each age group.

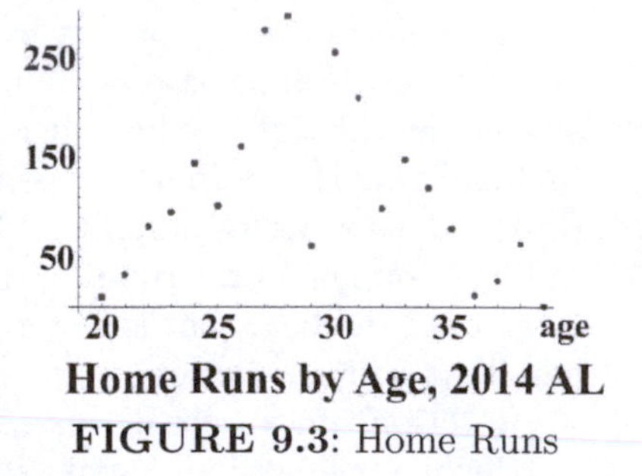

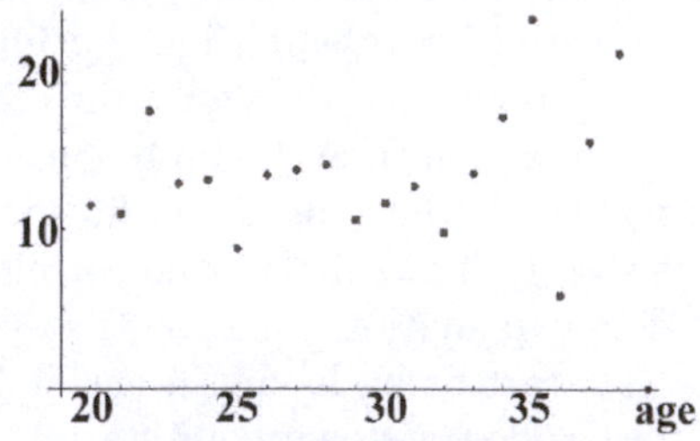

FIGURE 9.3: Home Runs

Figure 9.4 does this, and the result is quite different. Other than a rise in home run rate in the late 30s, the rates look fairly equal at all ages. Looking carefully at ages 23-28, you can see a general upward trend that is maintained. The lack of a clear downward trend following age 28 could be due to small sample sizes. You could reasonably object that Figure 9.4 only provides a snap shot of what happened in 2014. If we repeat the figure for 2015, will we find that 29-year-olds are the best? Perhaps this graph is more a function of selection bias (which players happen to be grouped together) than how players of all abilities age.

20
10
20 25 30 35 age

HR Rate by Age, 2014 AL

FIGURE 9.4: HR Rate

Let's look at a specific player. Ian Kinsler had his first full season in the major leagues at age 24, hitting 14 home runs. In succeeding years, his totals were 20, 18, 31, 9, 32, 19, 13, and 17. Kinsler only played 103 games in the year in which he hit 9 home runs, so a good analysis needs to account for injuries. He changed teams in 2014, so it would be good to account for the effects of ball parks. However, he hit the most home runs at ages 27 and 29, and there is a general drop before and after those years.

With this in mind, we could select a group of players with long careers and track their ups and downs. The selection bias here is that we are only looking at very good players, good enough to stay in the league for many years. Perhaps this group ages more gracefully than the average player. Still, this seems like a good way to go. Bradbury did a study like this and found peak ages of 28 for batting average, 32 for walks, 30 for home runs, and 29 for Linear Weights (see Example 7.5). Others have found a peak baseball age of 27, but in reading studies on this topic remember that different aspects of a sport may have different peak ages.

Personnel Decisions: Transfer Fees and Stars

A second issue with personnel implications is transferability of skills. The underlying issues here are individual skill and "fit" in a particular team's system. For example, if a running back ran for 1800 yards with one team and then changes teams, is it reasonable to expect another 1800-yard season? The book *The Success Equation* says no, that one of the few skills that holds up after changes of teams is punting.

How would you investigate this? Similar to aging, you need to choose a particular statistic to track. As far as collecting data, a simple idea would be to collect all before-and-after team changes for that stat. There are several factors you might adjust for. Style of the team is important: our running back moving to a pass-heavy team would not have the chance to duplicate his 1800-yard season. You could use yards per carry or percent of the team's rushing yards instead of total yards. In the year before becoming a free agent, players tend to have productive years, and then drop off after signing. Part of the drop-off is regression to the mean; our 1800-yard running back was likely to drop back closer to the league average if he stayed on his original team.

A third personnel issue has to do with team composition. In Chapter 8, we see that basketball teams with unequal scoring distributions tend to perform well. The theory that you need multiple stars to win in basketball is explored in the book *Scorecasting*. The short answer is "yes you do."

Anderson looks at this issue for soccer in *The Numbers Game*. As his measurable statistic (metric), he uses the Castrol Performance Index, which (like WAR for baseball) uses play-by-play data to try to measure the total contribution of a player to a team. He then looks at how teams' records depend on the value of the top players on the teams and on the bottom players on the teams. Not surprisingly, a team's record improves as its star player gets better, and also improves if its worst player gets better. The interesting result is that the value of the worst player has more of an effect on overall team performance than does the value of the star player. In this way, soccer seems to be a team sport in which the chain is only as strong as its weakest link.

To try a similar study, I took the eight position starters for each 2014 MLB team and computed the sums of the three highest and three lowest WAR values, as posted on baseball-reference.com. Figure 9.5 shows the scatter plot of the best WAR values against team wins. There is a mild trend for better teams to have higher top-three WAR values; the correlation is 0.53.

On the other hand, Figure 9.6 shows the scatter plot of the worst three WAR values against team wins. There is little difference visually between Figures 9.5 and 9.6, and the correlations of 0.53 and 0.49 confirm that the worst WAR values predict team wins as well as the best WAR values do.

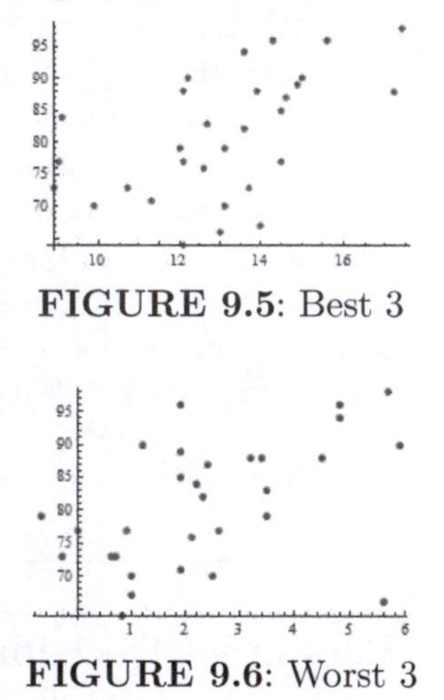

FIGURE 9.5: Best 3

FIGURE 9.6: Worst 3

A different version of the same study is to do a multiple regression, which produces the best-fit linear function $68.10 + 6.17b + 2.96w$ for wins, where b is the WAR value of the best player and w is the WAR value of the worst player. Since the coefficient of b is larger than w, we conclude that an increase in WAR value of the best player is more important than an increase in WAR value of the worst player. The "r-squared" value of the model is 0.49, indicating that 49% of the variation in wins can be explained by variations in WAR values (and that the model and wins have a correlation of 0.7). Pitchers were not included in the data. This gives some mild evidence that baseball is a star-oriented sport.

The conclusions here are dependent on the data that go into the study. If the player ratings are flawed, then any conclusions will be flawed. However, the Oakland A's and other low-budget sports teams have been using a "bottom-up" strategy of team building. They can't afford the big stars with the huge salaries, but they can build a strong team by making sure that even the worst players on the team are good. Getting a few young players to perform at high levels is their hope for gaining the boost that stars provide.

Exercises

9.1 Using $p(x) = 6 - 0.07x$, estimate the expected number of points for a football team starting at (a) its own 20 (b) its own 10. (c) At what position is the expectation 0? (d) Explain why this might be a fair place to start an overtime period.

9.2 Compute expected score values of going for it versus punting (assume a 34-yard net gain) on 4th and 2 at (a) your own 20 (b) your own 10. (c) How crazy is Kevin Kelley to never punt?

9.3 In example 9.2, the probability of making the first down in 0.53. How small would the probability have to be for the punt to have the larger expected value?

9.4 Suppose the result of an onside kick is for one team or the other to recover at the kicking team's 45. Find the break-even point, the probability of recovering the kick such that the onside kick has a positive expected value.

9.5 Suppose the result of a kickoff is for the receiving team to get the ball at its 35. Find the break-even point of an onside kick (see exercise 9.4), the probability of recovering the kick such that the onside kick has a larger expected value than a kickoff.

9.6 Suppose the options on an extra point are a sure 1 point or 2 points with probability 0.53. Which has the larger expected value?

9.7 Find the points value of a pass play from a team's own 20 to the 50. Show that the value is the same for a 30-yard gain from any location.

9.8 For the OneBase model, change the probabilities to 0.2 for a homer, 0.3 for a hit, and 0.5 for an out. (a) Write the transition matrix. (b) Write the expected runs vector for one batter. (c) Write the equations for total expected runs. (d) Solve those equations.

9.9 Find, as in Example 9.4, the expected run values for a runner on first with 1 out versus a runner on second with 2 outs,

9.10 Repeat example 9.5 with 1 out.

9.11 Suppose a tennis player wins points with probability 0.6 if the score is tied, 0.7 if ahead, and 0.4 if behind. If a game is won by the first player to win two points, set up the transition matrix for the states A (0-0), B (1-0), C (0-1), D (1-1), E (game won), F (game lost).

9.12 Given the win probabilities (half-inning averages, from baseball-reference.com) from August 3, 2015, compute the average win probabilities. (a) Atlanta 9, San Francisco 8: 54, 52, 30, 31, 21, 12, 5, 5, 3, 3, 2, 14, 19, 28, 27,18, 13, 50, 55, 100 (b) Texas 12, Houston 9: 39, 74, 83, 83, 83, 96, 94, 91, 93, 95, 96, 97, 96, 94, 97, 97, 100

9.13 A simple game for pitcher versus batter has a pitcher throwing a fastball or slider. When the batter guesses correctly, he bats .400 against the fastball and .300 against the slider. When the batter guesses incorrectly, he bats .200 against the slider and .350 against the fastball. Determine the best strategy for the pitcher and batter and the resulting batting average.

9.14 Repeat exercise 9.13 changing the .350 value to a more realistic .250.

9.15 The table gives minutes played and points scored by age in the 2014-15 NBA season. (a) Which age scored the most points? (b) Which age scored the most points per minute? (c) Discuss the optimal age for scoring points in the NBA.

	20	21	22	23	24	25	26	27
points	4533	11199	15984	19173	24540	23045	25335	18627
minutes	13646	28898	37387	50090	55979	51958	60456	46161

	28	29	30	31	32	33	34
points	21864	21914	13716	6817	8726	7275	8926
minutes	53652	51822	32524	17045	21826	17515	25154

9.16 In the 2014 American League season, the numbers of wins were {96, 71, 73, 85, 90, 70, 89, 98, 70, 84, 88, 87, 77, 67, 83}, the highest WAR values by team were {6, 4.8, 5.5, 7, 5.5, 6, 6.6, 7.9, 5.2, 4, 7.4, 6.4, 5, 7, 6} and the lowest WAR values were {.7, -1.4, -1.1, -1.4, -1.5, -.8, -1.2, -.3, -1, -1.4, -1, -1.5, -2, -1.9, -.5}. Create scatter plots and compute correlations of WAR values to wins.

9.17 In the 2014 American League season, the numbers of wins were {96, 71, 73, 85, 90, 70, 89, 98, 70, 84, 88, 87, 77, 67, 83}, the sums of the WAR values for the starters were {32.2, 19.4, 16.5, 18.1, 23.4, 14.1, 21.4, 33.2, 20.5, 13.3, 26.0, 19.7, 19.2, 14.9, 24.4} and the sums of the WAR values for the subs were {6.0, 2.0, -1.6, -1.6, 3.5, -0.2, -0.6, 2.3, 0.6, -0.5, 3.3, -0.7, -0.5, -2.4, 2.8}. Create scatter plots and compute correlations of WAR values to wins.

9.18 (T) (a) Discuss circumstances in which win probability added is a more informative football statistic than expected points added. (b) For evaluating the effect of a team's offense, discuss circumstances in which win probability added could unfairly treat mediocre and great teams differently.

9.19 (T) Discuss, in terms of expected points, whether a fumble is equally harmful at any location on the field. Give examples where win probability would change dramatically for a fumble at one location but not another.

9.20 (T) Discuss ways to implement a truly random strategy.

9.21 (T) Discuss why soccer goalies almost always dive left or right for penalty kicks, instead of playing the middle as often as calculations suggest is ideal.

9.22 (T) Discuss why penalty kick takers do not go high very often, even though high shots are almost never saved.

9.23 (T) Discuss how a team (sport of your choice) might implement a high-variance strategy. Does Kevin Kelley's strategy qualify?

9.24 (T) Discuss whether NBA teams who use the 2-for-1 strategy take lower-percentage shots.

9.25 (T) Discuss the differences between baseball, basketball, and soccer in terms of whether the quality of the best player is more important than the quality of the worst player.

9.26 (T) In *Analyzing Wimbledon*, it is reported that inexperienced players tend to hit conservative serves in pressure situations, whereas the top players show little or no change in service speed and placement. In game theory terms, discuss why it is important to maintain the same strategy in pressure situations.

9.27 (T) In *The 1984 Baseball Abstract*, Bill James notes that the 1983 National League East was highly compressed, meaning that the difference between first and last place was small. He also noted that the Cubs won fewer games in 1983 than their Pythagorean Method projection said they should have won. For these and other reasons, James called the Cubs a good long-shot bet, a team that could go from last to first. Explain why this was a reasonable prediction. (And, it turns out a good one: the 1984 Cubs did win the division.)

9.28 Ⓒ For the OneBase model, (a) write out the matrix S and (b) compute $(\text{I}-\text{S})^{-1}$.

9.29 Ⓒ Derive the formula $(\text{I-S})^{-1}\text{R}$ for the total expected runs in the OneBase model.

9.30 Ⓒ Compute the change in expected runs (use Tom Tango' expected runs matrix) for the following plays. (a) a leadoff single (b) a leadoff double (c) a single with one out and nobody on (d) a double with one out and nobody on (e) with one out and a runner on first, a single that advances the runner to third (f) with one out and a runner on first, a single that advances the runner to second (g) How much does the value of a single change with the number of outs? (h) How much more is a double worth than a single? (i) How many runs is it worth for a runner on first to get to third on a single?

9.31 Ⓒ Find the change in win expectancies for the situations in exercise 9.29 parts (a), (b), (e), and (f) assuming that (1) the score is tied in the bottom of the ninth (2) the team is down one run in the bottom of the ninth.

9.32 Ⓒ The probability of winning a service point in tennis is $ab + (1 - a)xy$, where a is the fraction of first serves that are in, b is the fraction of points won when the first serve is in, and x and y are the corresponding values for the second serve. (a) In *Analyzing Wimbledon*, $x = 0.86$ and $y = 0.51$. If the relationship between fraction in and fraction won is $b = 0.93 - 0.5a$, find the optimal first serve percentage. Compare to the actual average of 60%. (b) If only one serve is allowed, then the probability is ab. Find the optimal serve percentage. Is this closer to the actual first serve percentage of 60% or the second serve percentage of 86%? Discuss.

9.33 Ⓒ Team A leads by 3 points with 4 seconds left and fouls team B. Team B then fouls team A with 2 seconds left. Team B launches a 3-point shot at the buzzer. Assuming 70% free throw shooting, 30% 3-point shooting, and no offensive rebounds, what is the probability that team B ties the game? wins?

9.34 Ⓒ Reaction times and final times in the 2013 IAAF World Championship men's 100 meter sprint final are given. (a) Explore the importance of reaction times with a scatter plot and correlation. (b) For a given sprinter, lowering reaction time from 0.20 to 0.15 would improve the overall time by 0.05 s. Would that make a difference in placement for anybody? (c) Part (b) argues for a direct correlation between reaction time and final time. Reconcile this with the small correlation in part (a).

	1st	2nd	3rd	4th	5th	6th	7th	8th
reaction	0.163	0.163	0.157	0.186	0.142	0.158	0.154	0.177
final	9.77	9.85	9.95	9.98	9.98	10.04	10.06	10.21

9.35 Ⓒ Compare results from pro-football-reference.com's win probability calculator (use a Vegas line of 0, and 1st and ten) for the following situations. (a) trail by 1, 3 minutes remaining, 30 yards away; (b) trail by 1, 3 minutes remaining, 20 yards away; (c) trail by 1, 2 minutes remaining, 30 yards away. Then (d) estimate the value of 10 yards and one minute late in a game trailing by 1.

9.36 Ⓟ Expected points from different starting positions in football are nearly linear in distance from the goal line for the entire season. Investigate whether the linear relationship holds in special situations, such as the fourth quarter of a tight game or for a specific team.

9.37 Ⓟ Overall, NBA players have a higher shooting percentage in catch-and-shoot situations than for pull-up shots. Is this true of all teams? all players? For a potential game-winning shot, is the higher percentage play to go to the team's best shooter (who might have to take a dribble to get free) or kick it to a wide-open lesser shooter for a catch-and-shoot?

9.38 Ⓟ From game-by-game results, determine the correlation between yards rushing and winning. Would a high correlation prove that rushing creates wins? Determine the correlation between rushing yards in the first quarter and wins. Discuss whether rushing causes victories.

Further Reading

The Sloan Sports Analytics Conference videos are available online. "In-Game Innovations: Genius or Gimmick" is the panel discussion with Kevin Kelley and Bill James. Keating and Brenner's talk is "Giant Killers." Jeff van Gundy's comment is in 2015's "Innovators and Adopters."

More on Kevin Kelley can be found at http://grantland.com/features/grantland-channel-coach-never-punts/ Accessed 8-11-2015.

The books *Moneyball* and *Big Data baseball* give extended examples of the influence of analytics on baseball strategy.

In *The Hidden Game of Football*, Carroll, Palmer, and Thorn develop the concept of Win Probability, while exploring many of the ideas in this chapter.

"Monday Morning Math Modeling" by Hodds, Alcock, and Inglis in the February 2014 issue of the journal *Math Horizons* explores Belichick's strategy in the 2012 Super Bowl.

Expected runs matrices can be found at numerous sites and books, including Tom Tango's book *The Book* and website. Play-by-play win probabilities are part of the game descriptions at baseball-reference.com.

Win probabilities for football can be found at pro-football-reference (www.pro-football-reference.com/play-index/win_prob.cgi accessed 8-27-15).

Hot Stove Economics by Bradbury discusses peak ages in baseball.

Anderson and Sally's book *The Numbers Game* and Eastaway and Haigh's *How to Take a Penalty* discuss penalty kicks. Articles referenced are Chiappori, Levitt (yes, the *Freakonomics* co-author), and Groseclose: http://pricetheory.uchicago.edu/levitt/Papers/ChiapporiGrosecloseLevitt2002.pdf accessed 8-27-15 and William Spaniel's http://williamspaniel.com/2014/06/12/the-game-theory-of-soccer-penalty-kicks/ accessed 8-27-15.

An excellent model of aging in sports is in "Bridging Different Eras in Sports" by Berry, Reese, and Larkey.

Chapter 10

Big Data and Beyond

Introduction

A baseball batter hits a hard groundball into the hole. The shortstop, known as one of the best fielders in the game, backhands the ball and makes a spectacular jump-throw to first that the runner barely beats out. The batter is on first with a hit, but everyone applauds the all-out effort by the shortstop. That tale from the low-tech past may well be replaced, by the time you read this, with the following description. The batter hits a ground ball with exit velocity 81 mph. The shortstop has a very slow reaction time of 0.4 s, and only covers a distance of 8 feet before fielding the ball. The time between fielding the ball and getting off the jump throw is a very slow 1.3 s, and the throw has a low velocity of 58.3 mph. An average shortstop would have thrown out the runner 95.8% of the time, so the official scorer flashes "E6" indicating an error on the shortstop.

FIGURE 10.1: Jump

This chapter is about the brave new world of big data in sports. Massive data sets are coming online for all major sports. The challenge for teams will be to mine information that gives them competitive advantages. From a fan's perspective, this chapter is about the future as seen from 2015. Some of what is included here will be overly conservative; hopefully, little of it will be laughably off target. The following questions will be addressed. What types of new data are now available? What are some ways in which the presence of this data can help to measure previously unknowable quantities? What are some ways to visualize data that arrive in quantities that boggle the mind? Will big data affect the way that the casual athlete participates in sports?

Big Data Is Watching You

Modern sports contests are data collection extravaganzas. Arenas bristle with electronics, as cameras and sensors record every action on and off the field of play. Armies of computers sort and manipulate the data into forms that are useful for participants, media, and fans.

Through the All-Star break in 2015, Max Scherzer had thrown six different types of pitches (information at the Brooks Baseball website, from MLBAM data). His go-to pitch is the four-seam fastball, thrown 57% of the time; he only throws the two-seam sinker 1% of the time. How can we tell? The trajectories of his pitches are reconstructed by PITCHf/x, and different pitches move different distances in different directions. Scherzer's four-seamer averages 7.73 inches of horizontal movement and 7.92 inches of vertical movement, as compared to 9.66 and 7.31 inches, respectively, for his sinker. His average release point is more than two inches wider for his sinker than for his four-seamer. He throws his sinker three times as often against right-handed batters.

In 2014-15, Philadelphia's Michael Carter-Williams led the NBA with 100.5 touches per game, and also with 73.9 passes per game. Jimmy Butler of Chicago led the league with an average of 2.7 miles run per game. San Antonio's Patty Mills had the highest average speed at 4.8 mph. Among players who had at least 100 drives on the season, Toronto's James Johnson had the highest field goal percentage with 63.7% made. Detroit's Reggie Jackson led the league with 15.6 points per game created (scoring and assists) on drives. Minnesota's Ricky Rubio led the league with 1.3 free throw assists per game (passes to players who were fouled, missed the shot, but made at least one free throw). The Clippers' Chris Paul had 2.8 second assists (passes that led to assists) per game, and created 23.8 points per game from assists; both were league highs. Oklahoma City's Kevin Durant shot 43.5% on pull-up threes (at least one dribble), while Atlanta's Kyle Korver shot 50.4% on catch-and-shoot threes (no dribbles). (All numbers from nba.com.)

Most major soccer teams record locations of all players several times per second, from which velocity and acceleration can be computed. Soccer data can be hard to come by, but from www.dailymail.co.uk we learn that Lewis Holtby of Fulham ran 8.1 miles in a match against Liverpool on July 15, 2015. The day before, Jay Rodriguez of Southampton had reached a speed of 21.5 mph playing against Hull City. From the matchcenter at mlssoccer.com we learn that the New York Red Bulls connected on 48% of its crosses in its U.S. Open Cup match against the New York Cosmos. The Cosmos only completed 23% of its crosses and lost 4-1.

A graphic on the ESPN telecast of Wimbledon 2015 showed that Roger Federer never hit a first serve into Andy Murray's body in their semifinal match. Other graphics showed the positions on the court where the players hit returns of serve, with Federer much farther into the court than Murray. This data has not been released to the public for analysis.

At the 2015 MIT Sloan Sports Conference, Andrew Hawkins of the Cleveland Browns told a story about wide receivers wearing practice shirts wired to record speed and distance. The more competitive players would pull their chips out and throw them down the field to try to record the highest speed for the day. Coaches use practice data to judge whether players are working out at the proper level to prepare for the game while preventing injuries.

There is very little that an athlete does now that is not recorded and analyzed. The challenge is mining the data for useful information that can be clearly communicated to the coaches and athletes.

Example 10.1 Following up on Examples 9.4 and 9.5, discuss how data such as pitch movement, hitting charts, and running speeds could change the optimal strategy.

Solution Many responses are possible, including the following. The conclusion in Examples 9.4 and 9.5 is that the sacrifice bunt is an ineffective play. However, suppose that the pitcher throws a two-seam sinker that causes batters to hit ground balls 60% of the time. The batter hits ground balls 55% of the time, typically pulling them to one of two zones that the defense has covered by shifting its alignment. The chance of a double play is high, so the bunt might be a good play. On the other hand, suppose the pitcher throws a four-seam fastball and tends to give up fly balls. The batter hits line drives to all fields and has good speed. The chance of a double play is low, and the chance of a winning extra base hit is high, so the bunt would be a poor play.

A Theory of Everything

The Holy Grail of sports analytics has been a single number that measures the performance of every athlete in the sport. This formula would identify the best in the sport, would quantify how much money an athlete is worth in salary, and would quantify the fairness of a trade or the quality of a draft pick. A realistic appraisal of these goals should make you question whether such a formula could possibly exist. Nevertheless, the NFL Quarterback Rating and baseball's WAR and other rating systems are popular with the media and fans, if not the teams.

Baseball's WAR, soccer's Castrol Performance Index, football's Total QBR, and golf's Strokes Gained all share the same conceptual framework, which can be called "Value Added." First, organize the data so that the average outcome from any situation can be identified. Then, for every play for every athlete, compare the athlete's performance to the average performance, measured in units that are meaningful for the sport in question.

If a leftfielder hits 12 home runs in 132 games, compare that to the home runs you would expect his replacement to hit and quantify it in terms of wins above or below replacement level. Do the same for every fielding play and baserunning situation in the season, and you have WAR.

The Castrol Performance Index assigns points for passing accuracy, shooting accuracy, fouls given and taken, and so on, weighted based on the location of the play and how likely the play is to lead to or take away from a scoring chance. Total QBR takes into account the quarterback's ability to run, avoid sacks, deliver the ball to his wide receivers (even if the receiver then drops the ball), and so on, weighted by the importance of the play as determined by location on the field, score, and time remaining. In golf, Strokes Gained for a shot equals the difference between the expected score from the location before a shot and the expected score from the new location after the shot. Each type of stroke (putting, driving, and so on) can be evaluated separately or added together to get a total rating.

The different versions of WAR or Strokes Gained that you may read about differ in implementation of the system, but work from the same concept. The definition of a replacement player, the method for measuring the importance of a play, the granularity of the measurements (should a 92 mph line drive to left be measured against the same standard as a 93 mph line drive to left-center?), and the data used can all affect the final rating.

Example 10.2 A golfer plays a 480-yard par 4. He hits a 320-yard drive in the fairway, hits a 160-yard approach shot 22 feet from the pin, and two-putts for par. Determine Strokes Gained for each shot on this hole.
Solution A solution depends on several linear regressions on golf data, adapted from my *Golf By the Numbers*. From the approximation $f(x) = 2.77 + .0028x$, estimate the average score on a 480-yard par 4 as $f(480) = 4.114$. From the approximation $g(x) = 3.28 + .004x$, estimate the average score hitting the second shot from 160 yards away in the fairway as $g(160) = 3.92$. From the approximation $h(x) = 3.68 + .01x$, estimate the average score putting for birdie from 22 feet as $h(22) = 3.9$. The drive lowered the expected score from 4.114 to 3.92, so the drive saved 0.194 strokes. The approach shot saved $3.92 - 3.9 = 0.02$ strokes and the two-putt cost him $4 - 3.9 = 0.1$ strokes. With Strokes Gained values of $-.194$, $-.02$, and $.1$ (adopting the convention that better than average is under par and therefore negative), the player's overall rating is $-.114$, indicating that his 4 is .114 strokes better than the average (par) of 4.114.

Other examples of Value Added will be presented below.

Catch Me If You Can

While Big Data can help fine tune overall rating systems, the potential for changing the way sports are played and watched comes primarily from shedding light on aspects of the game that were previously unknowable. Analyses of the ability of baseball catchers to handle pitchers and control base runners have long been limited to vague quotes from players. Percentage of bases stolen has been used to evaluate catchers, even though everybody knew that pitchers have a large role in preventing or enabling runners to get good jumps. (In the three seasons from 2006 to 2008, only 10 runners tried to steal against Kenny Rogers; 8 were thrown out. Over a six year period Chris Young was run on 133 times with 126 bases stolen!) In all, we have been woefully ignorant of the so-called "tools of ignorance" employed by catchers.

Baseball Info Solutions (BIS) publishes *The Fielding Bible*. The 2015 edition shows how far we have come. BIS has defined a suite of useful measures of catching proficiency. On an attempted stolen base, **pop time** is the elapsed time from the ball hitting the catcher's mitt to the ball arriving at the fielder's glove for a tag. Then **tag time** is the elapsed time until the tag is made. Pop times generally range from 1.8 to 2.2 seconds. The percentage of base runners thrown out increases almost linearly from 36% for a pop time of 1.8 s to 25% for a pop time of 2.2 s with a right-handed pitcher on the mound (stealing quantity and quality are lower if the pitcher is lefthanded).

Tag time is used as a proxy for the catcher's throwing accuracy. Tag times for catchers range from 0.20 to 0.24 seconds, enough to matter but not as important as pop time. As noted, the pitcher plays a large role in base-running success. As one indicator of this, average caught stealing rates drop from 30% to 8% as the pitcher's delivery time increases from 1.3 to 1.65 seconds. And, of course, the speed of the runner is important. Caught stealing rates increase from 10% to 45% as the runner's time increases from 3.3 to 3.8 seconds. Additionally, the pitch types (caught stealing averages for fastball are 26-27%, versus 16% for curves) and pitch locations have measurable effects. There are other factors to be measured, such as ability to hold a runner close to first base, but the secrets of base-stealing are rapidly being revealed.

Example 10.3 Russell Martin had an average pop time in 2014 of 1.88 s. Estimate his caught stealing percentage. If the pitcher's delivery time is 1.4 s and the runner's time is 3.5 s, estimate the runner's chance of stealing a base.
Solution If caught stealing percentage is a linear function with $f(1.8) = 36$ and $f(2.2) = 25$, then $f(x) = 36 - \frac{36-25}{2.2-1.8}(x - 1.8)$ or $f(x) = 85.5 - 27.5x$. Based on pop time, Martin's caught stealing percentage should be $f(1.88) = 33.8$. (His actual percentage in 2014 was 32%.) For delivery time, the line through (1.3,30) and (1.65,8) is approximately $g(x) = 111.7 - 62.86x$ so a delivery time of 1.4 s should produce a percentage of $g(1.4) = 23.7$. For running time,

the line through (3.3,10) and (3.8,45) is $h(x) = 70x - 221$ so a time of 3.5 s should produce a percentage of $h(3.5) = 24\%$. Our three data points produce estimates of 33.8%, 23.7%, and 24%. While it is not clear how to average them (more research is needed!), a simple mean is 27.2%.

The next step here is to convert pop and tag times into runs saved (and, eventually, wins) by the catcher. A simple approach would be to create a formula into which you can plug in pop time and tag time and get a number. Given the available data, a better way of evaluating the catcher's performance is to do it play-by-play, taking into account delivery time, runner's speed, and whatever other information can be gathered. One issue is how many of these variables to include. In Example 10.3, taking the pitcher's delivery time and the runner's time into consideration, you could reasonably assign an expected percentage of 25.6% to this situation. If Martin throws out the runner, he gets credit for 74.4% of an out. Using the expected runs table discussed in Chapter 9, Martin could be given 74.4% of the difference in expected runs before the play and after the play.

Getting Framed

In the 1900s, fans would sometimes hear vaguely mystical statements about catchers being good receivers. That means catching the ball, right? It was hard to figure what the skill might be. Gradually, the skill of "framing" a pitch was clarified, and in the 2010s several researchers started quantifying the effect.

The idea of **framing** is that a catcher who lunges for a pitch sends a signal to the umpire that the pitch was not thrown well, and is probably outside of the strike zone. The same pitch caught smoothly without the lunge (its location clearly "framed" by the catcher's body and glove) is more likely to be called a strike. That seems reasonable, but the size of the effect was a big surprise to everyone.

The effects of pitch framing can be measured using the Value Added method. That is, we can map out the strike zone and record the locations of all pitches and the resulting umpire calls. A location in the center of the strike zone might be called a strike 100% of the time, whereas waist high on the outside corner might be a strike 65% of the time, and knee high on the inside corner might be a strike 40% of the time. Suppose Russell Martin catches a pitch in a location that is called a strike 40% of the time, and gets a strike call. Martin gets credit for +0.6 strikes better than average. Given data about batting averages for different strike counts, this can be converted to a fraction of a run saved. BIS's *Fielding Bible* states that, on average, the difference between a ball and strike call is about 0.12 runs. Add up all of the runs saved or lost for the entire season and you have an estimate of Martin's worth as a receiver.

How many runs in a year would you guess a good catcher saves? The first estimate (as told in *Big Data Baseball*, an excellent *Moneyball*-like chronicle of the resurrection of the Pittsburgh Pirates) was wild: the difference between good and bad catchers could be 300 runs in a season! Modern estimates divide that by a factor of 10, but 30 runs (about 3 wins) in a season is significant. This estimate does not include the psychological effect on the pitchers of being ahead in the count or being able to throw a borderline pitch with a full count.

The methodology described above is a first approximation that has already been adjusted. It makes little sense to give the catcher full credit or blame for a borderline pitch being called a ball or strike. Umpires have tendencies for being more or less generous with calls, a hitter who crowds the plate can affect an umpire's judgment, and a pitcher with good control is thought to get more calls than a wild pitcher (a paper by BIS's Rosales and Spratt finds that horizontal accuracy is more important than vertical accuracy). Credit and blame for balls and strikes calls must be divided among the catcher, umpire, pitcher, and batter.

Big data has revealed that pitch framing is real and an important skill that can be measured.

Anonymous Field Goal Kicking

The percentage of field goals made in the NFL has steadily risen from about 42% in 1952 to about 85% in 2012. All fans know how painful it is to see their kicker miss, particularly near the end of a game. This may disguise how infrequently NFL kickers miss. You might expect that with misses being a rarity it is hard to distinguish individual kickers. (Could you tell, without keeping score, the difference between free throw shooters who make 83 and 85 free throws, respectively, out of 100?) Research by Pasteur and Cunningham-Rhoads comes to this conclusion.

The structure of this study is to compute Value Added scores for each kicker. For each kick, compare the result to the expected make percentage for that type of kick. The word "type" requires definition. Which variables should be considered? You can argue that a large number of variables affect field goal percentage: distance, angle, wind, temperature, the quality of the defense, whether the kicker is tired or not, and so on. At some point, you have to say that the effect is not large enough to bother with. A linear regression can help determine which variables are useful and which are not: if the inclusion of a new variable only improves the total error by 0.01% it is not wise to retain the variable. In this study, only one of the above variables (angle) did not make the final model, although distance is far and away the best predictor of accuracy.

Example 10.4 Use the formula $p = \frac{1}{1 + e^{-5.8409+0.1078d}}$ for the probability of making a field goal of d yards to compare accuracy from (a) 30 yards, (b) 40 yards, (c) 50 yards.

Solution. (a) $\frac{1}{1 + e^{-5.8409+0.1078*30}} = 0.931$. (b) $\frac{1}{1 + e^{-5.8409+0.1078*40}} = 0.822$, only slightly lower. (c) $\frac{1}{1 + e^{-5.8409+0.1078*50}} = 0.611$, still well over 50%. The trend .931, .822, .611 is not linear, as the drop in accuracy from 40 yards to 50 yards is much larger than the drop in accuracy from 30 to 40 yards.

The distance-only model is reasonably accurate. The study takes advantage of this by stating the effects of other variables in terms of yards. (However, note that given the nonlinear nature of the probabilities in Example 10.3, the effect of an extra yard is much greater at 50 yards than at 40 yards.) The altitude of Denver subtracts about 3 yards, the difference between kicking against a good defense and a bad defense (as measured by points per game) can be 4 yards (surprising to me), the difference between hot and cold temperatures can be 6 yards, a 25-mph crosswind adds 7 yards, the playoffs add 4 yards, and the difference between the first attempt of the game and the fifth is 7 yards (another surprise).

So far, we have gained valuable information about how much certain factors affect field goal accuracy. However, what can be learned when the probabilities are used to evaluate kickers? The first conclusion is that field goal percentage does a poor job of rating kickers. The Raiders' Sebastian Janikowski did not make a (relatively) high percentage of kicks, but the offensively challenged Raiders often sent him out to try 55- and 60-yard kicks. Given the difficulty of his kicks, Janikowski rates as one of the best kickers of his era. That last statement should be read with suspicion. Is the rating persistent with Janikowski always high on the list, or is the rating essentially random with Janikowski sometimes floating to the top by luck? Irrespective of Janikowski's ability, the sad truth is that the correlation in performance of kickers from one year to the next was 0.01, essentially zero.

On a relative basis, then, the ranking of kickers has a large component of randomness. With few kicks being missed, a kick or two off the upright can drop a kicker well down the rankings. Plus, the kicker has no control over whether his team needs him frequently from short distances or infrequently from long distances.

On the Rebound

What is the best way to evaluate a rebounder in the NBA? Think about what can be misleading about total rebounds. The team's pace, the opponents' shooting percentage, the positioning of the players, the philosophy of the coach, and other factors affect the opportunity to gather rebounds. With player tracking that records the positions of the ball and players at all times, the possibilities for new rebounding metrics have exploded. The metrics discussed here come from a paper by Maheswaram, et.al., titled "The Three Dimensions of Rebounding."

The overarching concept is, again, Value Added: compare a player's rebounds to the expected number of rebounds for an average player in the same situation. The researchers divide the rebound into three stages. First is **positioning**: given where the players are when the shot is taken, what is the probability of getting the rebound? Second is **hustle**: between the time of the shot and the time of the ball becoming reboundable (reaching the 10-foot height level), how much has the player improved his or her chances of getting the rebound? Third is **conversion**: in what fraction of the player's rebound opportunities does the player end up with the rebound?

Imagine watching a replay from an overhead camera. Stop the action when a shot is taken. Some players are in the lane close to the basket, others are out at the three-point line. What percentage of time should each player get the rebound? The first information you need is where the rebound is likely to occur. Collect enough data and you can estimate the distribution of rebound locations for shots from different locations on the court. Now, imagine that a player is standing two feet away from the rim. This is good rebounding position, right? It depends: is this player the only one around, or are there other players in tight quarters banging for position? The second information you need is the amount of area that the player controls.

The second step involves constructing a **Voronoi diagram**. Figure 10.2 shows four potential rebounders around the basket. Player D seems to be in the best rebounding position, but this depends on where the ball bounces. For now, our goal is to divide up the court into the regions that each player controls. The idea is simple: for each spot on the court, whichever player is closest to the spot controls the spot.

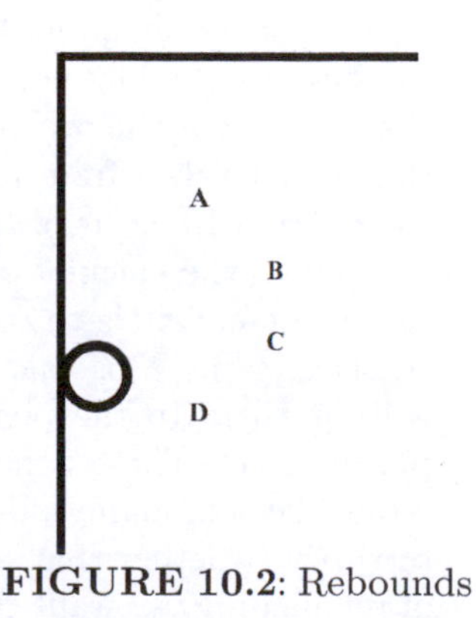

FIGURE 10.2: Rebounds

We use basic geometry to divide up the court. Start with two points, A and B. The dividing line between A and B is, in fact, a line. If you mentally sketch in a line that you think splits the difference between A and B, you will probably get very close to the dividing line. To be precise, though, start by connecting A and B with a line segment L. The dividing line we want is perpendicular to L and passes through the midpoint of L.

Figure 10.3 shows the six lines that divide two points at a time, labelled with the two points being separated. This creates a fairly large mess, but to get the final picture that we want (the Voronoi diagram) we simply erase a few of these line segments.

Many of the line segments in Figure 10.3 are unnecessary. For example, look at the line that goes just above the letter B in Figure 10.3. This is the dividing line for points A and C. However, in this part of the court, player C is not relevant. This region of the court is well above the line separating B and C; player B is much closer than player C, so we do not need to worry whether A or C is closer. Similarly, look at the portion of the AD line to the right of the mess of intersections. Clearly, the choice here is between players B and C, so the dividing line for players A and D is irrelevant and can be erased.

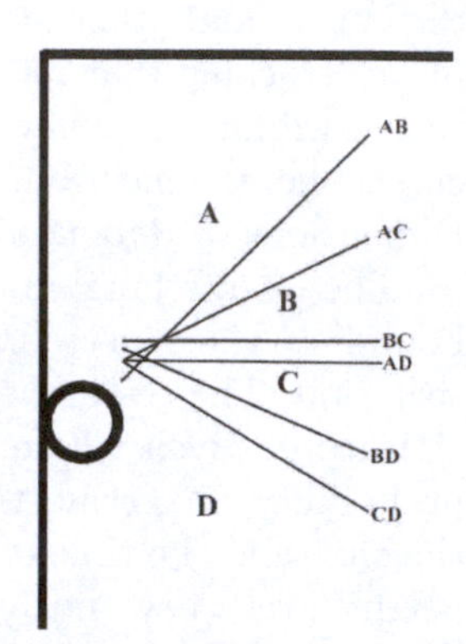

FIGURE 10.3: Dividers

Figure 10.4 shows the completed Voronoi diagram. Example 10.5 below gives you an idea of how a computer might construct this diagram.

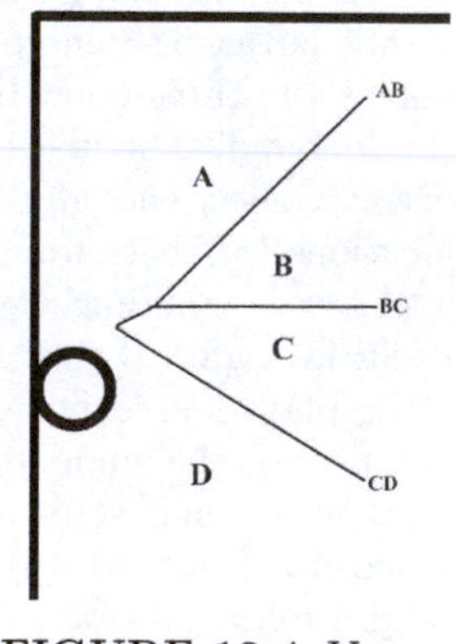

FIGURE 10.4: Voronoi

The Voronoi diagram is an integral part of the computation of the rebounding metrics of positioning, hustle, and conversion. For the positioning value, we take the Voronoi diagram and overlay the likelihood of a rebound coming into that region. For example, player D has a sizable chunk of area in Figure 10.4, but if the rebound is highly likely to occur in the upper portion of the diagram, then player D is not in the best rebounding position. The player's positioning value is the probability that the rebound (for a shot taken from a particular position on the court) will occur in the portion of the court that the player controls.

Repeat the calculation after a missed shot has hit the rim. The basis for the hustle metric is to compute the difference between positioning value and the new value. This calculation needs to be tweaked, however. Other players will crowd in to the picture, taking away territory originally controlled by players A-D. These other players will tend to go to open territory, so the larger the positioning value is for a player, the more likely that player is to lose territory (whether that player hustles or not). To be a different "dimension" of rebounding, we want the correlation of hustle and positioning to be zero, so the dependence of hustle on positioning needs to be removed. The final hustle value, then, is Value Added: the change in rebounding probability compared to the normal change in rebounding probability for a player with the given positioning value.

The conversion metric starts by identifying "opportunities." This is where a ball is at rebounding height in the player's Voronoi region. Conversion is the fraction of opportunities for which the player actually gets the rebound, above or below the average conversion rate for the given positioning.

Example 10.5 For players A (5,5), B (7,9), and C (6,7), (a) find an equation of the dividing line between A and B; (b) determine who is closest to the ball at (10,6).
Solution. (a) The segment between A and B has midpoint (6,7) and slope $\frac{9-5}{7-5} = 2$. The dividing line therefore has slope $-1/2$ (slopes of perpendicular lines multiply to -1). The line through (6,7) with slope $-1/2$ has equation $y = 7 - \frac{1}{2}(x - 6)$. (b) The distance between (10,6) and (5,5) is $\sqrt{(10-5)^2 + (6-5)^2} = \sqrt{26}$; The distance between (10,6) and (7,9) is $\sqrt{(10-7)^2 + (6-9)^2} = \sqrt{18}$; The distance between (10,6) and (6,7) is $\sqrt{(10-6)^2 + (6-7)^2} = \sqrt{17}$. Player C is closest.

It is not the intent of this section to claim that these three measures of rebounding are the proper way to evaluate rebounding. Instead, the intent is to give an example of how the new player tracking data can be used to calculate detailed aspects of the game.

Breaching the Convex Hull

In the 2014-15 regular season, Atlanta's Kyle Korver averaged 8 field goal attempts per game, of which nearly two-thirds were catch-and-shoot three pointers. What is the value of having a player whose primary offensive contribution is to wait to be passed the ball to shoot a three-pointer? Coaches talk about the importance of "floor spacers" whose excellent long-range shooting forces the defense to guard them closely. This spreads out the defense and allows other players to operate more effectively near the basket.

In *Basketball Analytics: Spatial Tracking*, Stephen Shea develops a way of quantifying the effects of floor spacers, and details the importance of spreading the floor in the 2014 NBA Finals. Shea uses a mathematical construct called a convex hull. Figure 10.5 shows two defensive alignments. Clearly, the defense on the bottom is more spread out than the one on top. The defense on top is packed into the lane, leaving any potential three point shooters open. The defense on the bottom is guarding the three point line carefully, and has left room for drives to the basket and post moves inside.

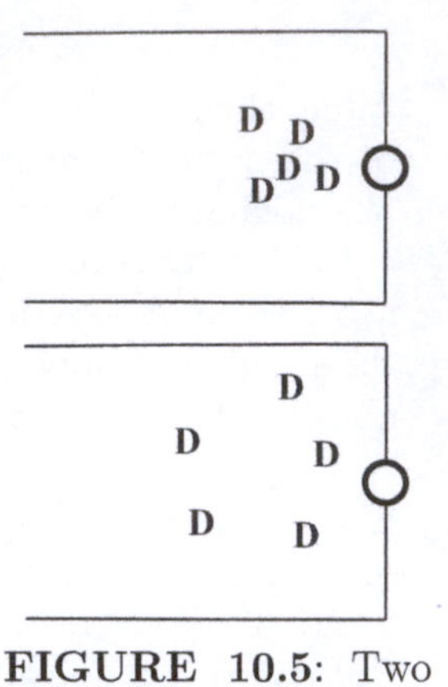

FIGURE 10.5: Two Defenses

The challenge is how to measure the spread of the defense. You could measure the "diameter" of the defense, the greatest distance between any two defenders. However, if one of the defenders on top moved away from the crowd, you would measure a larger diameter but you would not want to say that this defense is more spread out than the defense on the bottom.

Shea uses the area of the convex hull of the defenders to quantify spread. The concept is actually quite simple. Connect each pair of defenders with a line segment and then erase interior lines. The resulting polygon is the **convex hull**. The number of sides of the polygon is variable. Notice that one of the defenders in the top defense is on the interior of the convex hull, whereas the convex hull of the bottom defense is a pentagon. The area of the convex hull of the packed defense is clearly much smaller than that of the spread-out defense.

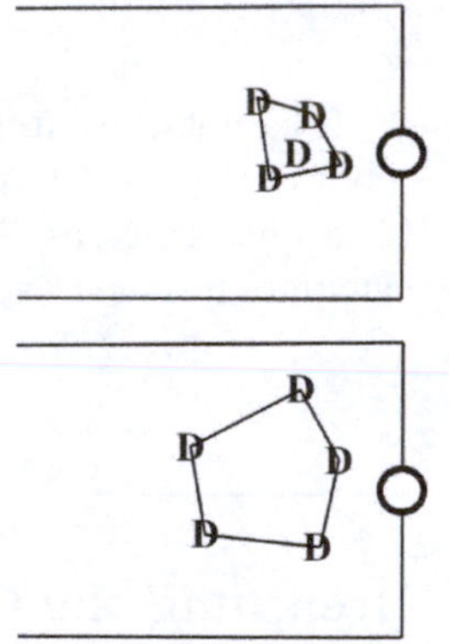

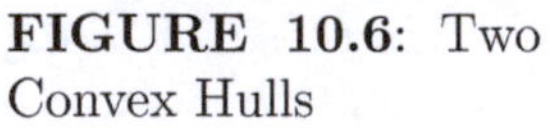

FIGURE 10.6: Two Convex Hulls

Shea computed areas of convex hulls of the San Antonio defense against Miami in the 2014 NBA Finals. The areas quantify how the San Antonio defense clogged the lanes in the first three games and shut down Miami's offense. However, in game four increased playing time for Miami floor spacers Ray Allen, Shane Battier, and others forced San Antonio to spread its defense. Miami's offensive efficiency improved, and the Heat came from behind to win the championship.

Example 10.6 Compute the area of the convex hull of the points A (1,3), B (4,2), C (3,6), and D (2,4).

Solution The convex hull is the triangle ABC (sketch a quick graph to convince yourself that this is true). We compute the area of the triangle ABC using Heron's formula, which only depends on the lengths of the sides. The distance between A and B is $a = \sqrt{3^2 + 1^2} = \sqrt{10}$, the distance between B and C is $b = \sqrt{1^2 + 4^2} = \sqrt{17}$, the distance between A and C is $c =$

$\sqrt{2^2 + 3^2} = \sqrt{13}$. Heron's formula gives the area as $\sqrt{s(s-a)(s-b)(s-c)}$ where $s = (a + b + c)/2$. In this case, the area equals 5.5.

Calculus Box: A Goal-Scoring Model

The various Value Added metrics that we have seen depend on accurate estimates of what is expected to happen. Expected values examined in this chapter have come from massive data sets giving reasonable empirical estimates. Other expectation calculations may come from a more sophisticated use of statistics than has been presented in this book. The next example is of that type.

In soccer and hockey, scores are low and goals can only be scored one at a time. These are indicators that a statistical distribution called the **Poisson distribution** may apply. The distribution function is $\frac{\lambda^n}{n!}e^{-\lambda}$, which gives the probability that n goals will be scored in one unit of time, if an average of λ goals are scored in that (arbitrary) unit of time.

Example 10.7 A soccer league has an average of 2.7 goals scored per game. Use the Poisson distribution to estimate the probability that (a) 4 goals are scored in a game; (b) 0 goals are scored in a game; (c) 1 goal is scored in the second half of a game that is tied 1-1 at half.
Solution The average for a game is $\lambda = 2.7$. (a) The probability of $n = 4$ goals scored is $\frac{2.7^4}{4!}e^{-2.7} \approx 0.149$, about a 15% chance. (b) The probability of $n = 0$ goals scored is $\frac{2.7^0}{0!}e^{-2.7} = e^{-2.7} \approx 0.067$, about a 7% chance. (c) The score is irrelevant. By the Poisson model, the probability of getting $n = 1$ goal in a period of time in which the average is $\lambda = 1.35$ goals is $\frac{1.35^1}{1!}e^{-1.35} \approx 0.350$, about a 35% chance.

Part (c) of Example 10.7 points out one of the important hypotheses underlying the Poisson distribution: the timing of goals is independent. Whether the score is 0-0 or 1-1 or 4-0, the probability of n more goals being scored is governed by the Poisson formula. You may think that the independence assumption is not valid. However, in *The Numbers Game* Chris Anderson describes research debunking the theory that teams are scored upon more often immediately after scoring a goal than at other times. In fact, the goals seem to come independently during the major portion of the game.

Figure 10.7 compares the actual number of goals scored in the 380 EPL games in 2014-15 to predictions of goals scored using a Poisson model with an average of $\lambda = 2.57$ goals per game (data from Soccer STATS). The match is quite good. The most significant mismatch is that fewer two-goal games were played than predicted. The statistical significance of this mismatch is explored in Example 10.8. Further details about the mismatch can be found in Figure 10.8. In the discussion of independence above, you may have thought that all bets are off near the ends of games. A team down 1-0 may throw caution to the wind trying to get the equalizing goal. While this might produce more 1-1 games, it could backfire and produce more 2-0 games. Similarly, in a 1-1 game will both teams play conservatively and take the point or will they push forward to get a winning goal? The results are given next.

Goals	Actual	Pred.
0	31	29
1	77	75
2	88	96
3	85	82
4	56	53
5	27	27
6	10	12
7 or more	6	6

FIGURE 10.7: EPL

Figure 10.8 breaks down the scoring into home and away goals. The 2014-15 average for home goals is $\lambda_1 = 1.47$ and the average for away goals is $\lambda_2 = 1.09$. The eight most common scores are shown. The match is good, but not as precise as in Figure 10.7. Given that there are many more possible outcomes with smaller sample size, this is not surprising. In terms of two-goal games, note the large shortfall in 1-1 games compared to the model prediction. This is the main contributor to the deficit in two-goal games seen in Figure 10.7. The numbers for 0-2 games are not shown, but there were 14 such games compared to a prediction of 17 games.

Home	Away	Actual	Pred.
1	0	40	43
1	1	37	47
2	0	37	32
0	1	37	32
2	1	35	35
0	0	31	29
1	2	26	26
2	2	22	19

FIGURE 10.8: EPL

Example 10.8 Compute the probability, for a Poisson process with $\lambda = 2.57$, that 2 goals would be scored in 88 or fewer games out of 380.

Solution Under the assumptions, the probability of scoring two goals is $\frac{2.57^2}{2!}e^{-2.57} \approx 0.2528$. Treating games as Bernoulli trials, the probability of getting 2 goals in exactly 88 out of 380 games is $\binom{380}{88}(.2528)^{88}(.7472)^{292} \approx 0.03$ and the probability of 88 times or fewer is the sum from 0 to 88 of $\binom{380}{n}(.2528)^n(.7472)^{380-n}$, which is approximately 0.187. An occurrence of 88 or fewer 2-goal games would occur by chance about 19% of the time under a Poisson model, so the largest deviation from predicted values in Figure 10.7 is not statistically significant.

Showing Hot and Cold

The collection and analysis of massive amounts of data is of little use unless interesting results are effectively communicated. Sophisticated computer graphics play an important role in clarifying patterns. Heat maps have become a common way of displaying data. They lose some impact in the black and white of this book, but some comments may be useful.

A heat map displays data, generally two-dimensional, using colors. The choice of color palette is entirely arbitrary, although when creating heat maps you should keep in mind that reds and oranges are normally associated with "hot." Here is a simple example.

Example 10.9 When the strike zone is divided into nine regions, a batter has the given batting averages in the regions. Display this as a heat map.

$$\begin{bmatrix} .310 & .350 & .290 \\ .280 & .320 & .270 \\ .260 & .260 & .220 \end{bmatrix}$$

Solution We convert each number to a color. In color, you could range from .220 in blue to .350 in red. For black and white, the range could be light gray for .220 and black for .350. This is shown in Figure 10.9a.

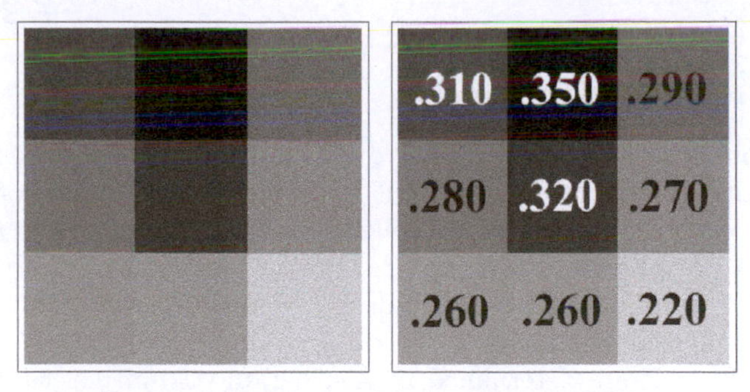

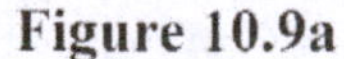
Figure 10.9a

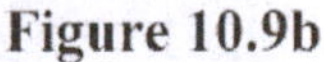
Figure 10.9b

FIGURE 10.9: Heat Maps for Example 10.9

For a web page, the point of a heat map may be to provide a colorful distraction, but if the point is to convey information labels are useful. If you came across Figure 10.9a out of context, could you tell that the darker regions are the batter's hot zones? In Figure 10.9b, the averages are included to remove ambiguity from the graphic. Notice also that the change from black to white font draws attention to the regions with averages over .300.

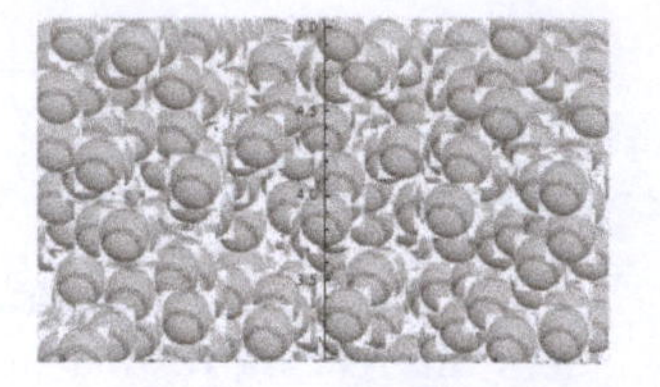

FIGURE 10.10: Pitches

A classic illustration of the power of graphics involves the display of pitch locations. *The New York Times* video "How Mariano Rivera Dominates Hitters" not only gives excellent information on Rivera's abilities, it also gives an invaluable lesson in communicating with graphics. The next two figures make the same point. Figure 10.10 displays the locations of three hundred simulated pitches. We display each pitch as a baseball. The "clever" use of the baseball image may be eye-catching, but little can be learned from the mess of balls.

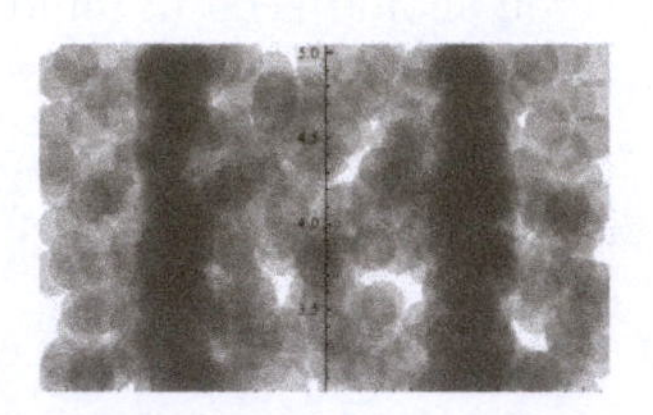

FIGURE 10.11: Pitches

For Figure 10.11, we tone down the cuteness and display each pitch as a see-through (opacity 0.2) disk. With the low opacity, you can see where most of the pitches end up. We can now learn that this pitcher threw a high percentage of pitches on the two corners of the plate. The pitcher's control and effectiveness are demonstrated by the graphic's extra dimension in which we can see the balls stacking up on the corners.

RIP to the RPI

The analytics movement in sports will have several indirect effects as quantitative thinking spreads throughout the system. An example of this involves Roanoke College, a Division III college with a long history of excellent soccer teams. Despite its history and the strong performance of its conference in the NCAA tournament, at-large bids to the tournament became increasingly scarce for members of the conference.

At-large bids are determined by a selection committee, which gives large consideration to the **RPI** (Rating Percentage Index). The RPI for a team is 25% of the team's winning percentage plus 50% of its opponents' winning percentage plus 25% of opponents' opponents' winning percentage. Roanoke coach Ryan Pflugrad thought to analyze the effect of Roanoke's large conference size (12 teams) on RPI.

Example 10.10 Compute the RPI for teams A and B: (a) A's record is 14-3, its opponents' record is 157-132, and its opponents' opponents' record is 1200-800; (b) B's record is 14-3, its opponents' record is 164-125, and its opponents' opponents' record is 1200-800.
Solution (a) Winning proportions are .8235 (14/17), .5433 (157/289), and .6000 (1200/2000) so A's RPI is $.25(.8235) + .5(.5433) + .25(.6) = .6215$.

(b) Winning proportions are .8235 (14/17), .5675 (164/289), and .6000 (1200/2000) so B's RPI is .25(.8235) + .5(.5675) + .25(.6) = .6336. Despite the identical 14-3 records and 60% opponents' opponents' records, B has a significantly higher RPI because its opponents won 7 more games than A's opponents, indicating that B played a tougher schedule.

Most likely, the outcome of Example 10.10 does not strike you as unreasonable. RPI rewards the team that played the tougher schedule. The lessons from Chapter 5 should make you a little suspicious of a system that only uses wins and losses, but the two components of RPI that attempt to quantify strength of schedule might ease that suspicion. Put into proper context, however, Example 10.10 actually reveals a major flaw in the RPI.

Consider two teams of equal ability and equal strength of schedule, except that team A plays in a 12-team conference and team B plays in an 8-team conference. To keep the comparison simple, suppose that both teams go 14-3, and that their opponents' opponents' records are identical 1200-800 marks. So, we restrict the difference in RPIs to the opponents' records. Team A plays 11 conference games and 6 out-of-conference games, while team B plays 7 conference games and 10 out-of-conference games. Suppose that A goes 10-1 in conference and 4-2 out of conference, and B goes 6-1 in conference and 8-2 out of conference.

Table 10.1 summarizes the records of A's and B's opponents. For both, the out-of-conference opponents (OC) win about 60% of the games: 61-41 for A's 6 opponents, 100-70 for B's 10 opponents (100-70 actually rounds to 59%). For both, the conference opponents win 60% of their out-of-conference games (C-OC): A's 11 opponents go 40-26 (6 games each), B's 7 opponents go 42-28 (10 games each). Finally, the conference schedules are equal in the sense that every team in each conference plays all other teams exactly once. In conference, then, A's 11 opponents play 10 games against non-A teams; in each game, one team (not A) wins and one team (not A) loses so the records sum to 55-55. Since A went 10-1 in conference, the other teams went 1-10 against A, making the total conference record 56-65 (shown in the C-C column). Similarly, B's conference opponents go 21-21 plus 1-6 for a total of 22-27.

TABLE 10.1: Records of Opponents

	OC	C-OC	C-C	Total
Team A	61-41	40-26	56-65	157-132
Team B	100-70	42-28	22-27	164-125

You have undoubtedly noticed that teams A and B in this scenario are the teams A and B in Example 10.10. The apparently tougher schedule of team B is simply an artifact of being in a smaller conference. We see that a team's excellent conference record harms the records of its opponents. If team A plays more conference games than B, A's strength of schedule is harmed more than B's. Example 10.10 shows that the difference between a 12-team conference and an 8-team conference is significant.

This calculation oversimplifies the effects of playing more or less conference games, but there is a clear lesson that the RPI penalizes teams in larger conferences. The punchline: Roanoke's conference, presented with the analytics argument, voted to split (for soccer) into two six-team divisions.

Blackbox Analytics

While it is true that computers merely follow instructions, when the instructions include ways to modify the instructions and evolve, amazing and unpredictable results can occur. Most of us have felt the wrath of a computer virus, a harmful example of an "autonomous" program, and you have probably heard of machine learning. We next take a quick look at neural nets, a popular type of machine learning.

Suppose that we want to use Four Factors to predict which team wins a basketball game. Recall that the four factors are shooting, turnovers, rebounds, and free throws. (As presented in Chapter 7, Four Factors was retrodictive; that is, we used the results of the game itself to retroactively evaluate who should have won the game. We could also use Four Factors to predict the outcome of a game, plugging in season averages for each team.) The question is how to combine the four inputs into a single number that predicts the outcome of the game.

A neural network typically consists of three layers: inputs, hidden, and outputs. We can think of having four inputs (the difference in the teams' shooting, and so on). For Example 7.7, the four inputs would be 0.032 (San Antonio's net shooting), −0.018 (turnovers), −0.92 (rebounds), and −0.116 (free throws). These four inputs are fed into the hidden layer.

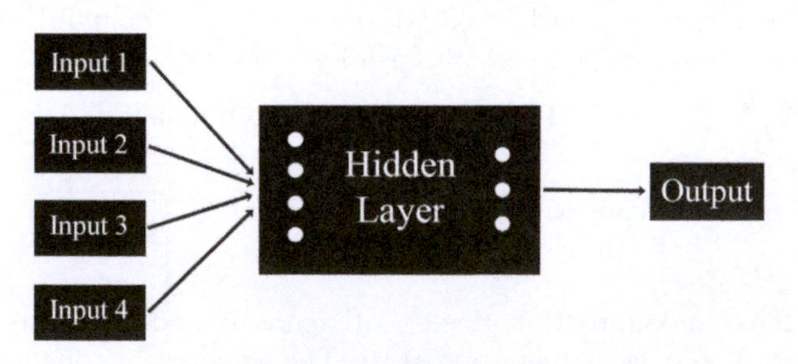

FIGURE 10.12: Neural Network

The general neural network in Figure 10.12 shows two columns of dots or nodes (the number of nodes is arbitrary) in the hidden layer. The idea is that each input could be entered into each node, and each node could compute a different function of the inputs. Outputs from the first set of nodes could be fed into a second column of nodes, with each of these nodes creating its own combination of its inputs. The final column of nodes could then be combined into the final output. The hidden layer of the network can be very complicated, such that it can be quite difficult to trace the path of a single input through all of the nodes in the hidden layer and determine its effect on the final output.

We have choices to make for the structure of the hidden layer. The simplest is to compute the weighted sum of the inputs and output "San Antonio wins" if the sum is positive, or "Los Angeles wins" if the sum is negative. For example, if the weights are 25, -10, 1, and 0.5, we compute $25(0.032) - 10(-0.018) + (-0.92) + 0.5(-0.116) = 0.002$. The positive value indicates that we predict a San Antonio win (which is correct; the Spurs won, 111-107). This is equivalent to a linear regression. We could use the small size of the output to say that the game was very close and output a probability. Using a sigmoid function, we could output $\frac{1}{1+e^{-10*0.002}} = 0.505$ and say that San Antonio had a 50.5% chance of winning the game.

Neural networks are modeled on the action of neurons which fire when a threshold is reached. With this in mind, we could compare each input to a threshold and send an output of 1 or -1 if the difference is significant enough. For example, a shooting difference of 0.032 might be highly indicative of a winning team, a rebounding difference of -0.92 indicative of a loser, and the turnover and free throw values too close to call. The output would be $1 + 0 - 1 + 0 = 0$, and we would not predict a winner. Or, each input could be plugged into a sigmoid function to produce a probability (e.g., a team that has a net shooting value of 0.032 wins 74% of the time) and the probabilities averaged.

Whichever choices we make for the structure of the neural network, there are constants to be determined (e.g., the weights 25, -10, 1, and 0.5) in the hidden part of the network. The magic of a neural network is that the network itself determines these values by training on large numbers of examples (e.g., for the current values of the constants, predict the San Antonio-Los Angeles game and 100 other games, and see how many predictions are right). The constants are tweaked to improve the predictions as much as possible. Practitioners generally use multiple training sets to try to avoid the neural network adjusting itself to oddities of the training set. A famous example involves a neural network that was supposed to learn to distinguish satellite pictures of tanks from pictures of civilian vehicles, but instead learned to distinguish sunny from overcast days because all of the pictures of tanks in the training set were taken on overcast days.

The programmer of a neural network does not have to know much about how the inputs relate to the output; the network figures that out. So, a belief that the Four Factors can be used to predict winners is enough to run a model. Depending on how many hidden layers are used and what type of thresholding is employed, the neural network may find patterns that would have eluded human observers. The patterns may be complicated enough that it is difficult to explain what is going on, even after the fact. This can be a distinct disadvantage of neural networks: if the analyst cannot explain how the network is getting its results, the player or manager may dismiss the results out of hand.

Neural networks have been applied to the complex problem of predicting the next baseball pitch. As such models increase in accuracy, you can imagine an arms race of sorts. Would it be unethical or illegal to transmit such predictions to a batter? Would the catcher and pitcher have a similar network telling them what the batter is likely to be looking for? The game theory discussion in Chapter 9 comes into play here, in that optimal strategies are supposed to be implemented randomly. This would foil any such pitch prediction algorithms. The future may be highly mathematical!

Neural networks and other machine learning techniques are powerful and flexible tools for discovering relationships in complex networks.

PeeWee Analytics

Figure 10.13 shows the trajectory and some data from a recent golf swing of mine at the driving range. My backswing is more extended than my downswing. The "ratio" of 4.6 measures tempo and equals the ratio of the time spent in the backswing to the downswing. Other data, recorded but not shown here, include a ball launch angle of 12.3°, a closed clubface angle of 3.9°, an attack angle of 0.9°, and a plane angle of 44°. Some of this data may create paralysis by analysis, but having a record of what my swing looked like when I was hitting the ball well could prove invaluable.

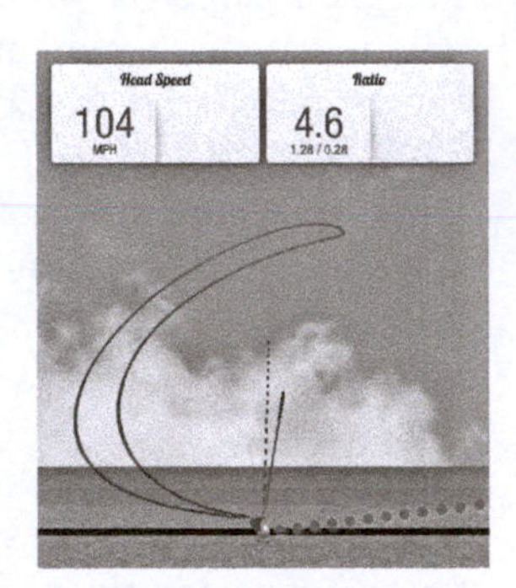

FIGURE 10.13: Drive

In addition, any golfer can now buy products that record the details of every shot on the course. In the clubhouse, traces of where every shot went (overlayed on a picture of the hole being played) can be printed out and statistical analyses of Strokes Gained putting, driving, and so on displayed.

All of the data collection and analysis tools discussed in this chapter, plus many more, will be available soon to coaches in 10-year-old recreation leagues of all sports. Whether this is a good or a bad development is debatable: I would have loved this when I was 10, but it probably would have done more for my development as a mathematician than for my development as an athlete.

Wearable Tech

Imagine a quarterback preparing for a game with a game simulator, somewhat like *Madden Football* viewed on your sunglasses. He sees the opponents' blitzes and zone defenses in realistic three-dimensional graphics that fully prepare him for the upcoming game. The quarterback coach reviews the quarterback's eye movements during the simulation and reminds him (again) that he needs to look off the linebackers on passes over the middle. The sensors in the quarterback's shoes record a left/right force imbalance that indicates that the quarterback's left knee is still sore. For the game, a patch will be applied to the knee that monitors the functioning of the knee and delivers a pain suppressant as needed.

The backup quarterback's simulator session did not go as well, as he fielded several phone calls on his simulation glasses. His jersey recorded an unusually rapid buildup of lactic acid, indicating that he is still out of shape. His hydration levels were also abnormal, possibly due to aftereffects of the party the night before. His eye movements and response times showed lingering effects of the concussion he had suffered in a previous game.

Does this sound like fiction? Perhaps by the time you read this, it will be commonplace. The scenarios required little imagination, as all of this technology exists in 2015.

The most reliable prediction for the future in sports is that technology will revolutionize every aspect of the athletes' and spectators' experience, in ways that we cannot imagine today.

Exercises

10.1 Discuss how data such as catch-and-shoot percentage, pull-up shooting percentage, drive percentage, and so on could affect defensive strategy in basketball.

10.2 Discuss how data such as miles run, speed, and acceleration could affect substitution strategy in soccer.

10.3 Discuss how charts showing where shots are hit from and to could affect service and return strategies in tennis.

10.4 Discuss how wearable tech in shoes, shirts, and so on, in practice could be used by coaches in making decisions about demoting starters or changing substitution patterns.

10.5 (a) Repeat Example 10.2 with the approach shot finishing 60 feet from the hole. Explain why the overall Strokes Gained is the same even though the approach shot and putt values have changed. (b) Compute Strokes Gained for each shot on a 380-yard par 4. A 300-yard drive in the fairway is followed by an 80-yard approach shot 20 feet from the hole, then two putts.

10.6 Estimate the caught stealing rate for a (a) pop time of 1.98 s; (b) delivery time of 1.3 s; (c) running time of 3.4 s. (d) Use your answers to (a)-(c) and Example 10.3 to estimate the effect of 0.1 s on the caught stealing rate.

10.7 (a) Use $h(x)$ from Example 10.3 to estimate the running time needed to have a steal rate of 80%. (b) For a pop time of 1.8 s and delivery time of 1.2 s, estimate the caught stealing rate. (c) For a pitcher/catcher duo with the values of part b, is your estimate from part a too high or too low?

10.8 At the *Fielding Bible's* rate of 0.12 runs per strike, how many extra strikes would a catcher need to frame to save his team (a) 30 runs? (b) 300 runs? (c) Do you think framing can save 300 runs in a season?

10.9 (a) Use the distance model from Example 10.4 to determine the distance at which NFL kickers would make 50% of their field goals. Does it seem realistic? (b) Repeat for 25%. (c) Repeat for 10%.

10.10 An NFL kicker who misses 3 or 4 field goals in a row is likely to be dismissed. Describe how this could cause the low persistence of 0.01 in Pasteur's field goal kicker metric.

10.11 Find an equation of the dividing line for each pair of points. (a) $(4,1)$ and $(4,-3)$; (b) $(4,1)$ and $(6,-1)$; (c) $(4,1)$ and $(6,5)$.

10.12 Sketch the Voronoi diagram for players at points $A(4,1)$, $B(4,-3)$, and $C(6,-1)$ (use exercise 10.11).

10.13 Determine which player is closest to the ball at point P, player $A(4,0)$, $B(6,-3)$, or $C(5,-2)$ (a) P is $(8,1)$ (b) P is $(2,-5)$.

10.14 Repeat Example 10.6 moving C to (3,7). Is the defense more spread out?

10.15 Compute the area of the convex hull of the points A (1,0), B (4,2), C (3,1), and D (5,1).

10.16 Draw a heat map for the batting averages shown: $\begin{bmatrix} .230 & .270 & .220 \\ .280 & .340 & .310 \\ .280 & .330 & .360 \end{bmatrix}$

10.17 Find heat map-like graphics in at least three different sports (links provided at the book's website). Critique the presentation of each. Is it clear which colors represent good play? Is it clear which regions have a large sample size for a "hot" or "cold" designation to be significant? Is it giving you information that you want?

10.18 Compute the RPIs of teams with the given record, opponents' record, and opponents' opponents' record. (a) 10-1, 58-63, 312-380 (b) 8-4, 80-64, 510-422

10.19 Keeping opponents' records the same, in Example 10.10 what would B's record have to be to make its RPI worse than A's?

10.20 For Table 10.1, we adjusted C-C to account for the effect of A's and B's records. We did not adjust OC or C-OC, however. (a) Explain why C-OC does not need to be adjusted. Speculate on whether C-OC would be better or worse if each team played more games. (b) To adjust OC, start with a 60% mark for each team's opponents, 57-38 for A and 96-64 for B. Add in the 2-4 mark against A and 2-8 against B and recompute the table. Does B's advantage over decrease?

10.21 (T) The various Value Added statistics depend on large data sets and calculations that the average fan could not do by hand. Discuss whether this reduces the reliability of the stat, or the charm of the stat, or the likelihood of the stat becoming accepted by fans.

10.22 Ⓣ Baseball's WAR statistics are value added "above replacement" players, while golf's Strokes Gained is strokes better or worse than average. Discuss which standard is more appropriate, the average value or the value of a replacement (potentially minor league) player.

10.23 Ⓣ Suppose that you hear a football analytics expert say that successful long passes are more dependent on the receiver than on the passer, while successful short passes are more dependent on the passer. Explain in football terms why this might be true. Discuss the evidence that might be collected to back up such a statement.

10.24 Ⓣ ESPN's Total QBR attempts to rate quarterbacks on all aspects of their play, including running and passing. Discuss whether the rating should depend on "leverage" (the importance of the play as measured by changes in win probabilities).

10.25 Ⓣ Discuss ways in which knowledge of delivery times and pop times could be used to decide whether to try to steal a base or not.

10.26 Ⓣ Find a video showing catchers framing pitches (e.g., http:// grantland.com /features/studying-art-pitch-framing-catchers-such-francisco-cervelli-chris- stewart-jose-molina-others/ accessed 9/11/2015) and describe what the catcher is doing.

10.27 Ⓣ It was reported that the field goal kicker evaluation metric has a persistence of 0.01. Discuss the meaning of this: field goal kicking at the NFL level is not consistent, the metric is flawed, or something else.

10.28 Ⓣ If RPI penalizes teams in larger conferences, should it be used to help rank teams for a tournament?

10.29 Ⓣ In basketball, is it easier to get an offensive rebound on a long or short shot? *Basketball on Paper* reports that NBA teams get offensive rebounds on 33% of 2-point shots and 31% of 3-point shots. Discuss the extent to which this answers the question posed.

10.30 Ⓒ (a) Show that the function p in Example 10.4 is decreasing and explain why this is important. Find limits of p as the distance goes to (b) 0; (c) infinity. (d) Which is these values is not realistic?

10.31 Ⓒ Suppose that the distribution of rebounds for a shot is inversely proportional to the distance from the rim for $1 < d < 15$. Compute a double integral to find the value of k in the distribution function $f(x, y) = k/\sqrt{x^2 + y^2}$ (assuming that the basket is at the origin and the foul line is at $x = 15$). If player A is at $(2, -1)$ and player B is at $(4, 1)$, find the positioning values for each player.

10.32 Ⓒ Compute the area of the convex hull of A (0,0), B (1,2), C (4,2), D (5,0) in two ways. (a) Show that the convex hull is a trapezoid and compute its area. (b) Find the areas of each of the four triangles formed and combine them in the appropriate way.

10.33 Ⓒ Repeat Example 10.7 for hockey's average of 5.32 goals per game.

10.34 Ⓒ NHL hockey games average 5.32 goals per game, so it has been said that the NHL is a "3-2 league." (a) Using the home team average of 2.78 goals per game and the road team average of 2.54 goals per game, compute the probability that a game ends 3-2. (b) Is 3-2 the most likely score?

10.35 Ⓟ Construct a neural net model for predicting basketball games. (See, for example, Loeffelholz et.al. in *JQAS* 2009.)

10.36 Ⓟ Research the methodology of *genetic algorithms* and construct a machine learning algorithm for predicting basketball games.

10.37 Ⓟ Find data on the number of baseball pitches of different types thrown by pitchers, and compute the pitch-type entropy for each pitcher. What can a batter expect from a high-entropy pitcher? Investigate whether high-entropy pitchers are as a group better, worse, or the same as low-entropy pitchers.

10.38 Ⓟ *The Hardball Times 2014 Baseball Annual* reports that pitcher's velocity in a game peaks in the first 20 pitches, then gradually declines for subsequent pitches. Investigate this claim.

Further Reading

A list of data-rich websites will be maintained at this book's web site. In 2015, the official league sites all have good information, and the reference.com sites (e.g., baseball-reference.com) are excellent.

Golf By the Numbers presents a development of Strokes Gained.

ESPN's explanation of its Total QBR system is at http://espn.go.com/ nfl/ story/_/id/6833215/explaining-statistics-total-quarterback-rating accessed 9-14-15.

Annual guides include *The Fielding Bible* from Baseball Info Systems, the *Football Outsiders Almanac*, Hardball Times' *Baseball Annual*, and guides from the Prospectus family (Baseball, Basketball and Hockey).

Journals include the online *Journal of Quantitative Analysis of Sports* and *European Journal of Sport Science*. The MIT Sloan Sports Analytics Conference posts numerous research articles and videos.

Severini's *Analytic Methods in Sports* gives an excellent overview of statistical methods applied to sports. Other excellent books about sports analytics include *Baseball Between the Numbers* by Baseball Prospectus, *The Sabermetric Revolution* by Baumer and Zimbalist, *Mathletics* by Winston, *Basketball on Paper* by Oliver, *Basketball Analytics Spatial Tracking* by Shea, *Analyzing Wimbledon* by Klaasen and Magnus, *Stumbling on Wins* by Berri and Schmidt, and *The Book* by Tango and Lichtman.

"Who Is Responsible for a Called Strike?" Joe Rosales, Scott Spratt, SSAC 2015 proceedings accessed 9-10-15

An article on using data analysis to predict baseball pitches: http://www.sporttechie.com /2014/03/03/applying-data-science-to-predict-mlb-pitching-patterns/ accessed 8-24-15

Answers and Selected Solutions

Chapter 1: Projectile Motion

1.1 We have $c = -9.8$, $v_0 = 0$ m/s and $p_0 = 100$ m. Then $v(t) = -9.8t$ m/s and $p(t) = 100 - 4.9t^2$. At $t = 1$, $v = -9.8$ m/s and $p = 95.1$ m. At $t = 2$, $v = -19.6$ m/s and $p = 80.4$ m. At impact, $p = 0$ m. This occurs at $t = \sqrt{100/4.9} \approx 4.52$ s, at which time $v \approx -44.3$ m/s.

1.2 (a) $-32\sqrt{2}$ ft/s (b) $-64\sqrt{2}$ ft/s (c) 2

1.3 96.1 m, 80.4 m, 55.9 m, 21.6 m; −4.9 m, −14.7 m, −24.5 m, −34.3 m; $9dp_1 = -44.1$ m, $11dp_1 = -53.9$ m; 3, 5, 7, 9, ...

1.4 (a) $t = 1$: $p = 6.1$ m, $v = -8.8$ m/s; $t = 2$: $p = -7.6$ m, $v = -18.6$ m/s (b) $h = 10.051$ m, $t = 1.534$ s, $v = -14.036$ m/s

1.5 (a) 14.51 ft/s (b) 0.907 s (c) 4 ft

1.6 (a) 80 ft/s (b) 100 ft

1.7 (a) good (if entry angle ok) (b) not a swish, goes off backboard

1.8 29.9 ft/s to 30.2 ft/s

1.9 24.941 and 24.938; minimum force

1.10 (a) hits ground before base (b)3.65°, 5.1 ft

1.11 (a) no, too long (b) 7.7° to 9.0° (c) 8.0° to 9.2°, small decrease

1.12 $y = 5.1$, no

1.13 (a) yes (b) no

1.14 108.2 mph, about 14% more

1.15 $k \approx 0.00304$, 112.5 mph

1.16 $k \approx 0.212$, 59.2 km/hr

1.17 (a) spin left, Magnus down, ball drops (b) spin down, Magnus right, ball moves right (c) spin up and right, Magnus up and left, ball goes higher and moves left (d) spin down and left, Magnus up and left, ball hooks left (e) spin left, Magnus up, ball goes higher

1.18 5.29 m/s, 10.1 m/s, 11.11 m/s, 11.63 m/s, 12.05 m/s, 12.19 m/s, 12.35 m/s, 12.19 m/s, 12.05 m/s, 12.05 m/s; 10 m/s, 11 m/s, 11.5 m/s, 12 m/s, 12.2 m/s, 12.3 m/s, 12.3 m/s, 12.2 m/s, 12.1 m/s, 12.1 m/s; 10 meters; 10 m/s^2, 1 m/s^2, .5 m/s^2, .5 m/s^2, .2 m/s^2, .1 m/s^2, 0 m/s^2, $-.1$ m/s^2, $-.1$ m/s^2, 0 m/s^2,

1.19 14.28 m/s, 13.49 m/s; fatigue; 13.89 m/s; 14.70 m/s, drafting; 7.01 m/s, 6.09 m/s; less friction, air drag; fatigue

1.20 (a) 10.38 m/s, 10.86 m/s, running start (b) 9.095 m/s, 9.091 m/s, slower runners in relay

1.21 7.01 m/s, 9.09 m/s, fatigue

1.22 men $19.69d^{-.13}$ women $18d^{-.133}$

1.23 157.5 mph

1.24 about 33% less

1.25 $\frac{30}{\pi}\omega$ rpm, 15 rev

1.26 Small area to produce spin, more pressure from fingers to roll ball

1.27 Since $x = ct$, can replace label x on x-axis with x/c on t-axis

1.28 Better testing for steroids, blood doping

1.29 (a) back, zero, forward (b) Magnus force pushes horizontally (c) Magnus force pushes ball forward at end

1.30 Longer contact with bat gives more spin gives more distance

1.31 Topspin up, Magnus down; curveball up right, Magnus down left

1.32 Ball rolls up instep so spin up right, Magnus up left; opposite

1.33 Swing plane tilts to right, spin up right, Magnus force up left

1.34 Less surface area on ball makes less drag.

1.35 (a) Changes direction too rapidly (b) Less drag but less Magnus force

1.36 Friction between feet and track, air drag

1.37 reduced air drag

1.38 more contact means more spin, more Magnus force, shots curve more

1.39 horizontal air drag is not balanced by gravity

1.40 Swimmers get speed by pulling their hands through the water. Pulling against a thick medium creates more force than against a thin medium.

1.41 The reactive force of the floor from the heavy soccer ball is passed on to the light racquetball, which bounces high.

1.42 For a putt moving up and right, the gravity vector pointed straight down is to the right of the velocity, accelerating the ball farther to the right. Moving downhill, gravity points to the left of the velocity vector, helping to straighten out the putt.

1.43 $v = c - ce^{-t} + v_0$, $p = ce^{-t} + (v_0 + c)t + p_0 - c$

1.44 slope -0.949, angle $-43.5°$, $x = 24.558$ ft, $0.447 > 0.346$

1.45 max at about $\omega = 1.2$ rad/s, limit is 0

1.47 126.7 ft/s vs 127.2 ft/s

1.49 0.389 s is less than 0.395 s; 93 mph pitch is effectively faster.

Chapter 2: Rotational Motion

2.1 (a) We have $\theta(0) = \theta_0$ rad and $\theta(T) = \theta_0 + 2\pi$ rad. Assuming that ω is constant, $\theta(t) = at + b$ for constants a and b to be determined. Then $\theta(0) = b$ and since $\theta(0) = \theta_0$, we have $b = \theta_0$. Then $\theta(T) = aT + \theta_0 = \theta_0 + 2\pi$ and so $aT = 2\pi$ and $a = \frac{2\pi}{T}$. This gives us $\theta(t) = \frac{2\pi}{T}t + \theta_0$. (b) We have $\theta(0) = \theta_0$ rad and $\theta(T) = \theta_0 - 2\pi$ rad. The steps in part (a) lead to $\theta(t) = -\frac{2\pi}{T}t + \theta_0$.

2.2 $60n$ rpm, $2\pi n$ rad/s

2.3 B: 20% more, C: 9.5% more ($\sqrt{2.4} \approx 1.095\sqrt{2}$)

2.4 (a) $t = \sqrt{\pi/2}$ s, $\omega = \sqrt{8\pi}$ rad/s (b) $t = \sqrt{2\pi/5}$ s, $\omega = \sqrt{10\pi}$ rad/s; 25% increase in α but 12% increase in ω

2.5 9.71 rad/s and 11.89 rad/s (22% more); not realistic

2.6 (a) $t = 1.944$ s, 15.7 rad/s (b) 5.2 rad/s

2.7 1.8 rad/s, less time to catch balance

2.8 436 lb

2.9 $120 < 144$ so defender has larger linear momentum; -1.6 ft/s

2.10 More than twice as fast; keep moving legs or push with arms

2.11 $\sqrt{27/2}$ m

2.12 $f(14)/f(13) \approx 1.239$, increase from 12 to 13 is more higher percentage increase than 13 to 14.

2.13 $144\rho\pi$ and $169\rho\pi$ (17% more); decrease density

2.14 $3/\alpha$ for pike, $4.61/\alpha$ (50% longer) for layout

2.15 $15/3.5 \approx 4.3$ times faster

2.16 $c \approx 2.96$

2.17 more weight near rotation point

2.18 longer rotation radius gives more clubhead/racket speed

2.19 cannot keep pushing/pulling legs and arms; simple approximation

2.20 Spiral has smaller MOI, making it easier to throw with high spin rate.

2.21 Harder to get high spin rate, but ball retains spin rate longer.

2.22 (a) PWS helps for serve, not for ground stroke; (b) PWS helps for ground stroke, not serve

2.23 chest is closer to center of mass than ankle so less torque

2.24 changing distance has more impact

2.25 increases MOI; from above, far left and right

2.26 reduces the MOI for flips

2.27 body rotates in the opposite direction

2.28 smaller MOI for a rotation, so easier

2.29 want smaller MOI when large rotation rates are good, e.g. swinging bat

2.30 (a) $c\pi[(121/192)L^3 - (121/192)24^3]$ (b) at $t = \sqrt{2\pi/\alpha}$ (e) for $30 < L < 35$, the graph of $48 - .34x$ is higher (f) constant α, shape of bat

2.31 $\frac{1}{4}c\pi[(R+2)^4 - R^4]$

2.32 approximately $8880c$

Chapter 3: Sports Illusions

3.1 In ft/s, $s = 75 * 528/360 = 110$ so $\theta' = \frac{2*110}{4+x^2}$ which has a maximum with $x = 0$ of 55 rad/s, well above the 3 rad/s limit.

3.2 (a) 6 ft/s (about 4 mph) (b) $\sqrt{84} \approx 9$ ft

3.3 (a) 77 rad/s (b) 9.9 ft (c) max rate of $\frac{30*154}{30^2} \approx 5$ rad/s, so no

3.4 (a) $\frac{300*264}{300^2} < 1$ rad/s (b) 88 ft

3.5 (a) 0.41 s (b) 0.20 s (c) 0.31 s (d) 0.23 s (e) 0.39 s

3.6 8.3%; 2.0%; 364 days

3.7 (a) 18.6° vs 15.8°, 11.9° vs 12.0° (better!) (b) 17.1° vs 13.9°, 10.0° vs 9.5°

3.8 (a) 1 yard, 9 inches (b) 3 yards, 9 inches

3.9 7.34 m, larger than record 6.16 m

3.10 position change is 2 m > 1.5 m

3.11 if D is at (0,0) and AR at (40,2) then (a) $0 < y < \frac{1}{20}x$ (b) $0 > y > \frac{1}{20}x$

3.12 (a) $0 < y < \frac{1}{40}x$ (b) $0 > y > \frac{1}{40}x$

3.13 at 10 m/s will move .5 m, could start on by .25 m, finish off by .25 m

3.14 1968 not special; Olympics may better reflect typical performance

3.15 The ball suddenly appears in an unexpected location.

3.16 no; he could have jumped gaze ahead to where ball meets bat

3.17 an unusual angle, few visual clues.

3.18 little experience deciphering pitch movement

3.19 $t = \sqrt{2d/g}$ is reaction time if stick drops d units

3.20 She anticipated the green light and has great timing.

3.21 pressure sensors on starting blocks; .1 s is less than best reaction time

3.22 nature: helpful physical traits; nurture: better training

3.23 long arms to grab ball, reach higher

3.24 moving heavy, light weights involve different muscles

3.25 transition from backswing to forward swing bends elbow, giving potential energy that increases racket speed

3.26 (b) because both players will be seen to have moved farther

3.28 flash-lag says the runner would appear farther ahead and therefore safe

3.29 the best jumpers had grown up using the flop

3.30 $\theta'(t) = \frac{1}{1+(x/2)^2} * (x/2)' = \frac{2s}{4+x^2}$

3.31 $\frac{Ls}{L^2+x^2}$

3.32 $w_x > 363$ rad/s or 3470 rpm

3.33 (c) 4.462 ft vs 4.570 ft (only 1 inch different) (d) 3.513 ft vs 3.789 ft (over 3 inches different)

3.34 $x = 17.7$ ft, the ball is still in the end zone!

3.35 $x = 24.0$ ft; no, still in the end zone

3.36 $x = \sqrt{27}$ ft, about 5 feet away

3.37 (a) 6 ft in from left post (b) 6 ft in from near post (c) 6 ft in from post unless kicker more than 6 ft in from post; then, even with kicker

---

Chapter 4: Collisions

4.1 (a) 220*18/320 = 12.375 mph (b) (220/32)*(18*5280/3600)/.2 = 907.5 lb (c) 907.5*2 = 1815 lb (d) 907.5*3/2 = 1361.25 lb

4.2 (a) defender wins (b) 6/7 mph (c) 707 lb (d) 1414 lb (e) 1060.5 lb

4.3 (a) 200 lb-s (b) 600 lb-s (c) 1800 lb-s (d) 800 lb-s

4.4 tendon (a), ball (a) since $x^2 > x^4$ for $0 < x < 1$

4.5 (a) .816 to .882; a little slow (b) .773 to .842; very slow (c) .728 to .762; very slow (d) .693 to .728; very slow

4.6 116 mph; no, the club moves forward

4.7 from height h, $v = \sqrt{2hg}$

4.8 (a) 115.7 mph (b) 114.7 mph (c) bat speed

4.9 (a) 144.7 mph (b) 166.9 mph

4.10 $(m_{bat}v_{bat} - m_{ball}((1 - \text{COR})v_{ball} + \text{COR}v_{bat}))/(m_{bat} + m_{ball})$

4.11 (a) decreases (b) increases if $m_{bat} > m_{ball}\text{COR}$

4.12 (a) 55.3 mph (b) 70.9 mph

4.13 To increase impact time, (a) pull hands back (b) bend knees and roll (c) start catching ball on way up (d) pull racket back

4.14 Greater impact time means sand absorbs more speed

4.15 looser strings give more power (trampoline effect), less control

4.16 decreases dramatically

4.17 off-center hit causes racket to twist; amount of twist affected by grip

4.18 a huge ball is hard to hit

4.19 B exits with more horizontal velocity so the angle is smaller

4.20 Yes if spin velocity high enough; topspin makes balls harder to reach because of higher horizontal velocity

4.21 lower; topspin; ball more likely to drop into basket

4.22 higher; backspin; volley can go higher/farther than expected

4.23 backspin; increase

4.24 w_{ball} up as v_{ball} increases; curveball; more backspin, more distance

4.25 club would dig into ground and ball would not go far; club hitting ball produces backspin which increases distance

4.26 larger COR; less friction; large rating

4.27 fingertips give more, creating more time and less force, making it easier to catch the ball

4.28 (a) impulse 133.3, force 666.6 (b) impulse 123.7, force 636.2

4.29 $\int_a^b (abt - at^2)dt = \frac{2}{3}bab^2/4$

4.31 (a) .927 (b) .768 (c) .940 (d) .845

4.32 $(m_{ball}(1 + \text{COR})(v_{bat} + v_{ball})/(m_{bat} + m_{ball})^2 > 0$ so a larger m_{bat} produces a larger w_{ball}

4.33 $-(m_{bat}(1 + \text{COR})(v_{bat} + v_{ball})/(m_{bat} + m_{ball})^2 < 0$ so a larger m_{ball} produces a smaller w_{ball}

4.34 (a) positive, larger bat gives larger bat speed (b) negative, larger ball gives smaller bat speed

4.35 increase in v_{bat} of 1 increases w_{ball} by 1.2

4.36 (a) $238m_{ball}/(m_{ball} + m_{bat})^2 > 0$ so larger bat gives more ball speed (b) $-238m_{bat}/(m_{ball} + m_{bat})^2 < 0$ so larger ball gives less ball speed

---

Chapter 5: Ratings Systems

5.1 (a) $\begin{bmatrix} 5 & -2 & -1 & -2 & 5 \\ -2 & 5 & -2 & -1 & -1 \\ -1 & -2 & 5 & -2 & 1 \\ -2 & -1 & -2 & 5 & -5 \end{bmatrix}$ (b) $\begin{bmatrix} 5 & -2 & -1 & -2 & 24 \\ -2 & 5 & -2 & -1 & 5 \\ -1 & -2 & 5 & -2 & 5 \\ -2 & -1 & -2 & 5 & -34 \end{bmatrix}$

(c) $\begin{bmatrix} 7 & -2 & -1 & -2 & 7/2 \\ -2 & 7 & -2 & -1 & 1/2 \\ -1 & -2 & 7 & -2 & 3/2 \\ -2 & -1 & -2 & 7 & -3/2 \end{bmatrix}$

5.2 Massey win A: 1.42, B: 0.67, C: 0.75, D: 0; Massey points A: 8.46, B: 6.5, C: 5.29, D: 0; Colley A: .775, B: .475, C: .525, D: .225; Massey points has B over C. Reasonable c values from 6 to 10. The points and win rankings are different.

5.3 (a) $\begin{bmatrix} 6 & -2 & -1 & -3 & 4 \\ -2 & 4 & -2 & 0 & 0 \\ -1 & -2 & 5 & -2 & 1 \\ -3 & 0 & -2 & 5 & -5 \end{bmatrix}$ (b) $\begin{bmatrix} 6 & -2 & -1 & -3 & 16 \\ -2 & 4 & -2 & 0 & 0 \\ -1 & -2 & 5 & -2 & 6 \\ -3 & 0 & -2 & 5 & -22 \end{bmatrix}$

(c) $\begin{bmatrix} 8 & -2 & -1 & -3 & 3 \\ -2 & 6 & -2 & 0 & 1 \\ -1 & -2 & 7 & -2 & 3/2 \\ -3 & 0 & -2 & 7 & -3/2 \end{bmatrix}$

5.4 Massey win A: 1.13, B: 0.97, C: 0.81, D: 0; Massey points A: 4.75, B: 4.31, C: 3.88, D: 0; Colley A: .67, B: .57, C: .54, D: .23; A, B, and C are now rated nearly the same, although the ranking remains A/B/C.

5.5 $5A_o - 2B_d - 3D_d = 72$, $4B_o - 2A_d - 2C_d = 44$, $4C_o - 2B_d - 2D_d = 40$, $5D_o - 3A_d - 2C_d = 40$, $-5A_d + 2B_o + 3D_o = 50$, $-4B_d + 2A_o + 2C_o = 44$, $-4C_d + 2B_o + 2D_o = 40$, $-5D_d + 3A_o + 2C_o = 62$, $A_d + B_d + C_d + D_d = 0$.

5.6 A: 13.82, B: 11.82, C: 9.82, D: 7.82; all have 7.82 added to old rating (so that the average rating is 10.82, the average number of runs scored)

5.7 (a) A 14, B 12 (b) A 15, C 11 (c) A 14, D 8 (d) B 11, C 9 (e) B 11, D 7 (f) C 10, D 8

5.8 No connection between AB and CD; no; (a) $\begin{bmatrix} 2 & -2 & 0 & 0 & 2 \\ -2 & 2 & 0 & 0 & -2 \\ 0 & 0 & 2 & -2 & 0 \\ 0 & 0 & -2 & 2 & 0 \end{bmatrix}$

(b) $\begin{bmatrix} 4 & -2 & 0 & 0 & 2 \\ -2 & 4 & 0 & 0 & 0 \\ 0 & 0 & 4 & -2 & 1 \\ 0 & 0 & -2 & 4 & 1 \end{bmatrix}$

5.9 (a) two rows of zeros, no column of -1s; $a - b = 1$, $c - d = 0$; A beats B, C and D equal (b) $a = 3/4$, $b = 1/4$, $c = 1/2$, $d = 1/2$; 2 wins boost A from 1/2 to 3/4, drop B to 1/4; C and D stay at 1/2

5.10 (a) $\begin{bmatrix} 3 & -2 & -1 & 0 & 1 \\ -2 & 2 & 0 & 0 & -2 \\ -1 & 0 & 3 & -2 & 1 \\ 0 & 0 & -2 & 2 & 0 \end{bmatrix}$ (b) $\begin{bmatrix} 5 & -2 & -1 & 0 & 3/2 \\ -2 & 4 & 0 & 0 & 0 \\ -1 & 0 & 5 & -2 & 3/2 \\ 0 & 0 & -2 & 4 & 1 \end{bmatrix}$ (c) C,D should rate higher than A,B

5.11 (a) C(0), D(0), A(−1), B(−2) (b) C(.63), D(.57), A(.53), B(.27); Colley gave C more credit for beating A than Massey

5.12 Scoring: Baylor, TCU, Marshall, Oregon, Ohio State (Marshall had bad SOS; Georgia was 8th, had good SOS); Defense: Ole Miss, Stanford, LSU, Alabama, Memphis (Memphis had bad SOS)

5.13 $\begin{bmatrix} 3 & 0 & 0 & 0 & 0 & -1 & -1 & -1 & 34 \\ 0 & 3 & 0 & 0 & -1 & 0 & -1 & -1 & 12 \\ 0 & 0 & 3 & 0 & -1 & -1 & 0-1 & -3 & \\ 0 & 0 & 0 & 3 & -1 & -1 & -1 & 0 & 4 \\ 0 & -1 & -1 & -1 & 3 & 0 & 0 & 0 & -10 \\ -1 & 0 & -1 & -1 & 0 & 3 & 0 & 0 & 3 \\ -1 & -1 & 0 & -1 & 0 & 0 & 3 & 0 & -20 \\ -1 & -1 & -1 & 0 & 0 & 0 & 0 & 3 & -20 \end{bmatrix}$ A(14.2), B(−3.3), C(0.6), D(4.2); BR better because in general home teams are better than road teams so BR has better SOS; DR plays AH, best team in league; AR does not

5.14 Column sums are 2, 2, 2, ..., n so $2r_1 + 2r_2 + ... + 2r_n = n$ or $r_1 + r_2 + ... + r_n = n/2$ and $(r_1 + r_2 + ... + r_n)/n = 1/2$.

5.15 change is $k(s_1 - m_1) + k(s_2 - m_2) = k(s_1 - m_1 + (1 - s_1) - (1 - m_1) = 0$

5.16 (a) 2802.4 and 2597.6 (b) 2792.4 and 2607.6

5.17 $\frac{1}{1+10^{-n}} = \frac{10^n}{10^n+1}$ which equals 10^n times $\frac{1}{1+10^n}$

5.18 2605 and 2595 and then 2599.9 and 2600.1; not the same.

5.19 should stay 7; increases to 7.1; to 7.3; weird

5.20 (a) 125.8 and 84.2 (b) 93.8 and 116.2

5.21 (a) change 15.9 or 16.1 (b) change 15.6 or 16.4

5.24 weight recent games more than early games

5.25 more teams get recognition, leaving a team out is more harmful, voters need to know more

5.26 (a) rating (b) ranking (c) ranking (d) ranking

5.28 where/when games are played, injuries, new starters

5.29 wins; right

5.30 Elo has equal numerical changes, but higher percentage change for underdog

5.31 opponents' ratings are part of equations

5.32 no, all numbers in equations double; yes, more games with which to change Elo ratings

5.34
$$\begin{bmatrix} 38 & 0 & 10 & 0 & 0 & -18 & 0 & -30 \\ 0 & 24 & 0 & 10 & -18 & 0 & -16 & 0 \\ 10 & 0 & 24 & 0 & 0 & -16 & 0 & -18 \\ 0 & 10 & 0 & 38 & -30 & 0 & -18 & 0 \\ 0 & -18 & 0 & -30 & 39 & 1 & 11 & 1 \\ -18 & 0 & -16 & 0 & 1 & 25 & 1 & 11 \\ 0 & -16 & 0 & -18 & 11 & 1 & 25 & 1 \\ -30 & 0 & -18 & 0 & 1 & 11 & 1 & 39 \end{bmatrix} \begin{bmatrix} 634 \\ 356 \\ 372 \\ 430 \\ -458 \\ -400 \\ -325 \\ -606 \end{bmatrix}$$

5.35 Sum of rows is zero, so the rows are dependent. If A plays B, then B plays A; if A wins by 4, then B loses by 4; the sums are zero.

5.37 (a) $0.734^5 = 0.213$ (b) 0.212

5.38 (a) schedule a (b) schedule b (c) 0.88 versus 0.94, schedule a

Chapter 6: Voting Systems

6.1 (a) Yes, Bradford would still win with 900+315=1215 points, to McCoy's 798+288=1086 and Tebow's 927+207=1134, but Tebow would finish second. (b) With $x = 20$, Tebow gets 20*309+414+234=6828 points to Bradford's 6826. (c) In 2008, Bradford would have had 300+315+196=811 votes to McCoy's 784 and Tebow's 750.

6.2 No, if e.g. he receives no 2nd or 3rd place votes.

6.3 (a) Alexander 10 2nd place votes for 48 points, Manning 30 2nd place votes for 56 points. (b) Alexander 10 2nd place votes for 48 points, Brady 30 2nd palce votes for 50 votes.

6.4 Rodriguez 7, Martinez 4; Martinez was a pitcher.

6.5 Anchorage dropped from 23 to 22 votes. New information.

6.6 (a) Lillehammer (b) Anchorage

6.7 Anchorage beats Lillehammer, 28 to 56.

6.8 (a) A with 8 (b) B with 21 (c) C with 11 (d) no (e) B with 16

6.9 (a) C (b) A (c) A (d) C last place in a majority of lists, but wins plurality.

6.10 (a) C (b) C (c) C (d) B, violates IIA

6.11 (a) 8 A/B/C, 7 B/C/A, 6 C/B/A: A plurality, B Condorcet (b) 10 A/B/C, 5 B/C/A, 4 C/B/A: B Borda, A Condorcet (c) same as (b) with two approved (d) same as (b) with points 3-2-1

6.12 Higher rated team always wins.

6.13 (a) A wins instead (b) C wins (c) violates IIA

6.14 (a) Number of first place votes (b) Order of elimination (c) Borda yes, others no

6.15 Votes differ by m (number of voters) for each candidate. 13 A/B/C, 4 B/A/C, 5 B/C/A

6.16 (a) yes (b) yes (c) no (d) no

6.17 In 6.1, change 2 of 11 voters in last column to order B/C/A: B wins.

6.18 Plurality-no, more first place votes helps, Borda-no, more points helps, approval-no, more votes helps, range-no, more points helps

6.19 (a) A with 3 (b) A with 22 (c) 4th highest is 2 (d) B beats C, 5 to 4 (e) A wins, 7 to 8 (f) A:2,1,1,3,2,3,5; B:3,3,2,1,1,5,3; C:4,2,3,2,3,1,1 (g) medians 2,3; C beats A, 16 to 17, B is third (h) Third place to first place is unreasonable.

6.20 (a) [1,0,0,0,0,0] (b) [10,7,5,3,1,0] (c) [7,4,3,2,1,0]

6.21 (a) $x > 5$ (b) $x < 5$ (c) never

6.22 (a) B gets one of A's two links and one of D's three links

6.23 (a) link to A from B 1/2 the time, always from C, from D 1/3 of the time

$$\text{(b)} \begin{bmatrix} -1 & 1/2 & 1 & 1/4 & 0 & 0 \\ 1/2 & -1 & 0 & 1/4 & 0 & 0 \\ 1/2 & 0 & -1 & 1/4 & 0 & 0 \\ 0 & 1/2 & 0 & -1 & 0 & 0 \\ 0 & 0 & 0 & 1/4 & -1 & 0 \end{bmatrix} \quad \text{(c)} \begin{bmatrix} -1 & 1/2 & 1 & 1/4 & 1/5 & 0 \\ 1/2 & -1 & 0 & 1/4 & 1/5 & 0 \\ 1/2 & 0 & -1 & 1/4 & 1/5 & 0 \\ 0 & 1/2 & 0 & -1 & 1/5 & 0 \\ 0 & 0 & 0 & 1/4 & -4/5 & 0 \end{bmatrix}$$

$$\text{(d)} \begin{bmatrix} a \\ b \\ c \\ d \\ e \end{bmatrix} = .85 \begin{bmatrix} 0 & 1/2 & 1 & 1/4 & 1/5 \\ 1/2 & 0 & 0 & 1/4 & 1/5 \\ 1/2 & 0 & 0 & 1/4 & 1/5 \\ 0 & 1/2 & 0 & 0 & 1/5 \\ 0 & 0 & 0 & 1/4 & 1/5 \end{bmatrix} \begin{bmatrix} a \\ b \\ c \\ d \\ e \end{bmatrix} + .15 \begin{bmatrix} 1/2 \\ 1/4 \\ 1/4 \\ 0 \\ 0 \end{bmatrix}$$

6.24 (a) 0.268 (b) 0.467 (c) 0.070 (d) 0.232 (e) 0.398

6.25 answers vary

6.26 Votes for a given player would be unaffected by votes for or against others. The negative player is replaced by B on enough ballots, the positive player replaces A on enough ballots.

6.27 answers vary

6.28 A beats B, B beats C, C beats A; equal but different

6.29 (a) e.g., Condorcet winner or number of 2nd place votes (b) e.g., Condorcet winner or number of 1st place votes

6.30 Yes, but the winner cannot be part of the Condorcet cycle.

6.31 (a) False - might not have majority of 1st place votes (b) True - will beat everyone (c) False - might not have enough points (d) False - might not have enough points

6.32 Place B last; only approve of A; 4 stars to A, 1 to others; all are subject to strategic voting

6.33 The slime mold changes behavior when 3rd alternative presented.

6.34 The existence of medium size shouldn't affect choice of large or small.

6.35 answers vary

6.36 A majority could prefer A over B, a (different) majority prefer B over C, and a majority prefer C over A.

6.37 answers vary

6.38 (a) 61 (b) 2 (c) 81

6.39 answers vary

6.40 (a) Voters can't determine who is best (b) The more voters, the more the randomness evens out.

6.41 (a) Give strong candidate the worst rating. (b) One biased or extreme judge in each direction gets ignored. (c) Yes, if too extreme or biased; yes, if only mimicking other judges.

6.42 One avoids negatives, the other maximizes positives.

6.43 An arbitrary choice that makes the results look good.

6.44 9-seeds win more 1st rounds, but 11-seeds win more 2nd rounds.

6.45 Suppose that A and B have the same number of wins, and that A beat B. In their other games, then, B won one more game than A, and therefore beat somebody (call it C) that A did not. The Condorcet cycle is A beat B, B beat C, and C beat A.

6.46 (a) Let A have the most wins (or tied for the most). For any other team B, either A beat B or B beat A. If B beat A, then in games against other teams A won more games than B, so there was at least one team C such that A beat C and C beat A. (b) If there is a complete Condorcet cycle (A beats B beats C ... beats A) then everybody wins a medal.

6.47 At least 3.

6.48 Ranking A first puts you left of the 25, so you prefer B to C. Ranking C first puts you right of the 45, so you prefer B to A. If a majority prefers A to B, the order A/B/C has a majority of votes, so A beats C. For a majority to prefer C to B, the order C/B/A must have a majority, so C beats A. If neither of these occur, B defeats both A and C.

6.49 The only orders are A/B and B/A, one of which has a majority, which also wins plurality and the Borda count. The majority winner might not be rated highly enough or approved of by enough voters to win.

6.50 Each wins with probability 2/3.

6.51 Yes, if there are 5 votes currently.

6.52 A: 0.228, B: 0.261, C: 0.163, D: 0.196, E: 0.152

6.53 A: 0.196, B: 0.204, C: 0.225, D: 0.189, E: 0.186

6.54 $log_2(1) = 0$, $log_2(2) = 1$, $log_2(3) = 1.58 \uparrow 2$, $log_2(4) = 2$, $log_2(5) = 2.32 \uparrow 3$ and so on. If w is the expected number of wins, w equals $7 - g$ where g is the rounded log value.

Chapter 7: Saber- and Other Metrics

7.1 The Cubs' winning proportion is predicted to be $701^2/(701^2 + 719^2) = .487$. Multiplying by 162 games, the expected number of wins rounds to 79.

7.2 Bad luck, better than their record showed.

7.3 93 > 82; good luck winning close games; no

7.4 exactly 85; very different; good luck didn't persist

7.5 .553, .573, .474, .409, .390; Los Angeles; yes; Pythagorean better for 3 of 5

7.6 Oakland −11 in 2014: 0, 2, −3, −4, −6, −1, −3, 8, −5 (usually negative); St. Louis 7 in 2014: −4, −5, 2, −5, 2, −5, 0, 0, 7, 1, 2 (average of 0)

7.7 Cleveland 0.271, Golden State 0.146, large Cleveland advantage

7.8 rapid decrease since 2007

7.9 walks steady decline since 2009; home runs decline since 2000; hits big decreases 2006-2010, 2014

7.10 slow increase with yearly ups and downs

7.11 rushing up and down, no trend; first downs increase since 2010

7.12 increase since 2004-05

7.13 field goals increase until 2009-10, then decrease; free throw small variations, no trend

7.14 (a) better offense 28-17, defense 29-16 (b) offense 28-17, defense 24-21 (c) offense 28-17, defense 28-17

7.15 .978; 534 (60 less); Hardy

7.16 .971 is lower; Aparicio made 100 more plays; number of ground balls hit by Baltimore opponents

7.17 0.44 pt/min and 0.39 pt/min; group scored more with Bogut; quality of opponents

7.18 (a) $209/150 = 1.39$ (b) 0.72 (c) 0.81 (d) 0.94

7.19 (a) 1.32 (b) 1.07 (c) 0.92 (d) 0.98

7.20 Cabrera linear weights 64.4 (Trout 55.1) and wOBA .457 (Trout .425) the better hitter

7.21 (a) Warriors 0.5 (+.040), 0.123 (+.013), 0.256 (−.004), 0.227 (+.89) won three, tied one (b) Warriors 0.51 (+.080), 0.091 (−.073), 0.149 (−.184), 0.212 (−.117) won two decisively, lost two decisively

7.22 Ovechkin: 326, 467, 16.1; Kane: 166, 220, 11.0

7.23 (a) Meredith 87.7, Jurgensen 84.5 (b) 1966: 64.2, 2014: 87.1

7.24 53.66, 151.67

7.25 158.33; Romo's stats in exercise 7.24 with 250 yards instead of 218

7.26 $0.78 * 1780 = 1388$ close to 1383

7.27 $0.831 * 4.2 = 3.49$ close to 3.44

7.28 $c = 1$: $x - (x - a) = a$; $c = 0$: $x - 0 = x$; smaller

7.29 relief pitching, clutch hitting, several blow-out games; continues if clutch hitting and pitching are repeatable skills

7.30 more playing time for players who walk often; if all teams get walks, no advantage gained

7.31 since OBP was more predictive of success, it is likely that OBP was undervalued

7.32 Points for a basketball sub who scores against the other team's subs. Points per minute when all players get similar playing times

7.33 a persistent statistic is predictive of the future and indicates a stable skill

7.34 putting has large element of luck, under pressure putting efficiency can decrease

7.35 it may measure the role of the player or team strategy more than a skill

7.36 winning depends on offense and defense of players not under the pitcher's control

7.37 Range Factor gives credit to fast defenders who make many plays, does not over-penalize errors

7.38 depends on quality of teammates and opposition, and the scoring pace of the teams

7.39 both give more credit to 3-pointers than 2-pointers, TSP accounts for scoring by free throws

7.40 each offensive rebound gives the team an extra shot for its one possession

7.41 the other team gets offensive rebounds, some shots go out of bounds without a rebound

7.42 it limits the impact of an extreme stat in a game

7.43 the relationships among the variables could change with new strategies, rules, players

7.44 trades and injuries make a team better or worse; these factors would not be different for even- and odd-numbered games

7.45 praise follows an unusually good play, yelling after an unusually bad play; the next plays are likely to be more normal. Yelling seems to work.

7.46 (a) a worse score (b) a worse game by regression to the mean

7.47 short passes are automatic completions and may not help the team score

7.48 1.81

7.49 (a) 3.01 (b) 14.51; larger exponents for higher scoring games

7.50 (a) 2.03 for points (b) 1.26 for points; less accurate because of increased importance of one goal

7.51 Rothman 0.1552 worse than Pythagorean 0.1439

7.52 $w_y = \frac{-2x^2y}{(x^2+y^2)^2}$

7.53 NFL $x = y = 361$ and $a = 3/01$ gives $0.5 + 0.00208(x - y)$; NBA $x = y = 8201$ and $a = 14.51$ gives $0.5 + 0.000442(x - y)$; NHL $x = y = 218$ and $a = 2.03$ gives $0.5 + 0.00233(x - y)$; coefficients are higher

7.54 (a) 2004 more linear (b) both very scattered (c) 2004 more linear

7.55 (a),(c) both very scattered (d) has tighter grouping than (b)

7.56 (a) has tighter grouping than (b); (c), (d) both very scattered

7.57 (a) 0.22 (b) 0.71 (c) 0.50 (d) 0.76

7.58 (a) 0.020 (b) 0.055; neither correlation is high but possession is higher

7.59 (a) 61.7 (b) 73.1 (c) 74.7 (d) 76.6 (e) 84.1; passing more often and more efficiently

7.60 −143.33+5.4136shots−0.008124possess+1.1038pass+3.1596fouls' shots and passing important, possession not, more fouls go with increased scoring

7.61 $4x - 3.3$; data is from a parabola $y = x^2$ not a line

---

Chapter 8: Randomness in Sports

8.1 The means and standard deviations for the American League East, Central, West, and National East, Central, and West are, respectively: ALE - 82.2, 9.31; ALC - 81.4, 9.29; ALW - 82.0, 13.10. The largest mean (best division) is the East. The smallest standard deviation (most balance) is the Central.

8.2 (a) 13 (b) 2 (c) 0

8.3 (a) B (b) B (c) A

8.4 (a) peak 30-35, equal numbers to the left and right but more spread out to the left (b) biggest peak 40-45, spread out to the left with another peak 20-25; Patriots had a wider spread

8.5 (a) nearly uniform distribution (b) somewhat normal distribution, too many values 15-20. The NBA seems to be normally distributed around the 50-50 mark, the NFL has more good/bad teams

8.6 won 2 versus spread; mean 0.2, standard deviation 9.3

8.7 84%

8.8 (a) 0.154 (b) $\sigma \approx 5$, 16%

8.9 (a) 0.026 (b) $\sigma = 4$, 16%

8.10 (a) 0.00012 (b) 0.046 (c) more likely

8.11 (a) 0.00058 (b) 0.0059 (c) very unlikely (1 in 170)

8.12 (a) 91 (b) $\sigma \approx 6.3$, 0.5 (c) 0.16

8.13 (a) 0.5301 and 0.53 (b) 0.5309 and 0.53 (c) 0.5444 and 0.53 (d) 0.6 and 0.6 (e) 0.5990 and 0.6

8.14 (a) Gini 0.44, entropy 1.276 (b) Gini 0.04, entropy 1.606

8.15 (a) Gini 0.37, entropy 2.050 (b) Gini 0.06, entropy 2.298

8.16 max 0.9, min 0

8.17 0.471

8.18 (a) 0.16 (b) 0.025 (c) Smaller σ so fewer large values

8.19 means: 148.3, 148.1, 156.6, 149.9, 146.1, 149.7, 147.6, 144.5; sigmas: 1.99, 1.98, 2.59, 1.86, 1.70, 2.51, 0.80, 1.03; no, but the courses change; 3 of the last 4 have been low; yes, the last two especially; based on the standard deviations, runners are better in general

8.20 0.027; probably not

8.21 independence: mean 11, variance 4.7; (a) 12 runs, fine (b) 7 runs, borderline (2 sigmas below mean)

8.22 (a) $\mu = 6$, $\sigma = 1$, 7 runs is acceptable (b) $\mu = 9.4$, $\sigma = 1.8$, 11 runs is acceptable

8.23 $\mu = 9$, $\sigma = 1.5$, 5 runs is too few; 0.002

8.24 $\mu = 207$, $\sigma = 9.7$, 206 is acceptable. Streak lengths are close to expectation. No evidence. If $p = 159/451$, the probability of 6 in a row is 0.0019. Given enough time, unusual things occur.

8.25 (a) $1 - 0.65^4 = 0.82$ (b) $1 - 0.65^5 = .88$

8.26 (a) 0.9375 (b) $0.9375^{56} = 0.027$

8.27 (a) 0.424 (b) 0.331

8.28 $(1-.84^{56})^{25}$ is the probability of failing 25 times to get a 56-game streak. 0.001436 vs 0.001437

8.29 a probability of 2 cannot be correct; $1 - (1 - .001)^{2000} = 0.865$

8.30 Kershaw .252 and 2.56 in 2014, .264 and 3.06 in 2013; Maddux .254 and 2.42 in 1994, .272 and 3.07 in 1993. BABIPs all better than average, Maddux least lucky in 1993.

8.31 Gwynn .388 and .364, well above average; Thome .353 and .319, above average

8.32 smaller; the closer p is to 1 the less chance of failures; also, $p(1-p)$ has a max value at $p = .5$

8.33 Improved technology can extend the performance limits.

8.34 balance fair competition with having recognizable stars

8.35 analytics has shown that batting average is not the best stat, so it is not selected for as rigidly

8.36 Boggs had to contend with better bullpens, more travel, but Williams had worse conditions

8.37 At possible last games (4-7), the 2-3-2 and 2-2-1-1-1 models have the same split of home and road games. The 4-3 model starts with a 4-0 split instead of 2-2; the 2-1-2-2-2 model starts with a 3-1 split.

8.38 Players who break serve may be better players, and more likely to hold serve for that reason.

8.39 at each step, H and T are equally likely; the runs test tests the pattern of streaks against expected patterns

8.40 the feeling of being unstoppable and the game becoming easy with success automatic; answers vary

8.41 (a) too many H's (b) no long streaks (c) no streaks of length 1

8.44 could be coincidence, but humans pay attention to round numbers

8.45 batters who hit line drives have higher BABIP, pitchers are affected by quality of defense (which does not change)

8.46 not informative, since Clemson was also unbeaten in its other uniforms

8.47 (a) peaks 95-100, 100-105; symmetric (b) peaks 90-95, 100-105; bell shape (c) peak at 2, long tail to the right (d) peaks at 1,2, shorter tail to the right

8.48 (a) decreasing ($1/x$ like) from 92 with 2-3 to 55 with 4-5 to 28 with 12-13 and so on (data from baseball-reference.com) (b) decreasing from 96 with 2 to 78 with 3 to 17 with 7 and so on

8.49 0.6827; this is the 68% of the empirical rule

8.50 (a) $20x^{-2}$ (b) $20e^{-x}$

8.51 about 3.9 points

8.52 8(b) 0.179 with 55, 0.131 without, averages 16%; 9(b) 0.189 with 76, 0.131 without, averages 16%

8.53 $\binom{20}{12} p^{12}(1-p)^8$; $p^{11}(1-p)^7(12(1-p) - 8p = 0$ so $p = 0.6$

8.54 1/3

8.55 If $f(p) = -p\ln p - (1-p)\ln(1-p)$, $f'(p) = 0$ if $p = 0.5$. $\sum_{i=0}^{n} -(1/n)\ln(1/n) = \ln(n)$

8.56 0.530 (2010), 0.523 (2000), 0.567 (1990), 0.492 (1980); no trends apparent

8.57 (a) both the same: 5.876 (b) 2-3-2: 5.786, 2-2-1-1-1: 5.723 (shorter)

8.58 36, since 33 at bats with at least 11 hits for a .200 hitter has probability 0.051.

8.59 (a) 0.09 (b) 0.051 (almost significant)

8.60 (a) 0.739 (b) 0.742

8.61 The probability of getting 4 or less from a binomial variable with parameters $n = 26$ and $p = .54$ is .000056.

Chapter 9: Sports Strategies

9.1 (a) $p(80) = 6 - .07(80) = 0.4$ (b) $p(90) = 6 - .07(90) = -0.3$ (c) $p(x) = 0$ if $x = 6/0.07 \approx 86$, at own 14. (d) Each team equally likely to score.

9.2 (a) -1.727 (go for it), -2.22 (punt) (b) -2.427 (go for it), -2.92 (punt)

9.3 $p = .42$

9.4 $p = .57$

9.5 $p = .28$

9.6 2-point conversion, $1.06 > 1$

9.7 $p(50) - p(80) = -.07(50) - -.07(80) = 2.1$ points; $p(x-30) - p(x) = -.07(x-30) + .07x = 2.1$ points

9.8 (a) $\begin{bmatrix} .2 & .3 & .5 & 0 & 0 \\ .2 & .3 & 0 & .5 & 0 \\ 0 & 0 & .2 & .3 & .5 \\ 0 & 0 & .2 & .3 & .5 \\ 0 & 0 & 0 & 0 & 1 \end{bmatrix}$ (b) $[.2, .7, .2, .7]$ (c) $a = .2 + .2a + .3b + .5c$, $b = .7 + .2a + .3b + .5d$, $c = .2 + .2c + .3d$, $d = .7 + .2c + .3d$ (d) 1.55, 2.3, 0.7, 1.2

9.9 .573 versus .344

9.10 .637 versus .610

9.11 $\begin{bmatrix} 0 & .6 & .4 & 0 & 0 & 0 \\ 0 & 0 & 0 & .3 & .7 & 0 \\ 0 & 0 & 0 & .4 & 0 & .6 \\ 0 & 0 & 0 & 0 & .6 & .4 \\ 0 & 0 & 0 & 0 & 1 & 0 \\ 0 & 0 & 0 & 0 & 0 & 1 \end{bmatrix}$

9.12 (a) 27.1% (b) 88.7%

9.13 The batter does worse against the slider, whether guessing fastball or slider. Since the pitcher's slider strategy dominates the fastball strategy, he should always throw the slider. Knowing this, the batter should always guess slider and will bat .300.

9.14 The strategies are mixed (random). For the batter, $.4p + .25(1-p) = .2p + .3(1-p)$ gives $p = .2$: batter guess fastball 20%, slider 80%. For the pitcher, $.4p + .2(1-p) = .25p + .3(1-p)$ gives $p = .4$: pitcher throws fastball 40%, slider 60%.

9.15 (a) 26 (b) 25 (c) 24-26

9.16 The correlation of best and wins is 0.39, and the correlation of worst and wins is 0.50.

9.17 The correlation of starters and wins is 0.75 and the correlation of subs and wins is 0.56.

9.18 (a) For events in close games, win probability may be more important. (b) Mediocre teams play many close games, with many opportunities to gain win probability points.

9.19 The expected points lost are the same at all locations. With 10 seconds left, a fumble at the 50 could be less harmful than one near the goal line.

9.20 Flip a coin or use a roulette wheel.

9.21 Tradition and potential ridicule for standing still.

9.22 There is a risk of missing by kicking too high.

9.23 Kelley's strategy is high variance in the sense that more points are likely to be scored.

9.24 They get the same shots as a team that lets the shot clock run down.

9.25 Individual players can dominate basketball and baseball more than high-level soccer.

9.26 The optimal strategy is a random mix of different types of serves, which is independent of the situation.

9.28 (a) $S = \begin{bmatrix} .1 & .2 & .7 & 0 \\ .1 & .2 & 0 & .7 \\ 0 & 0 & .1 & .2 \\ 0 & 0 & .1 & .2 \end{bmatrix}$ (b) $(I - S)^{-1} = \begin{bmatrix} 1.143 & .286 & .943 & .486 \\ .143 & 1.286 & .243 & 1.186 \\ 0 & 0 & 1.143 & .286 \\ 0 & 0 & .143 & 1.286 \end{bmatrix}$

9.29 The equations translate as $x = R + Sx$ where x is the 4x1 vector of unknowns a, b, c, and d. Then $Ix - Sx = R$ and $x = (I - S)^{-1}R$.

9.30 (a) .398 (b) .634 (c) .276 (d) .428 (e) .670 (f) .398 (g) drops .398 to .276 to .134 (h) about 50% more (i) .272 (.331 with 0 outs, .072 with 2 outs)

9.31 (1) (a) 0.081 (b) 0.173 (e) 0.192 (f) 0.074 (2) (a) 0.137 (b) 0.243 (e) 0.256 (f) 0.129

9.32 (a) About 49%, less than the actual value of 60% (b) 93%, closer to the second serve percentage. A change in rules would eliminate the first serve, not the second serve.

9.33 Tie: 12.7%, Win: 8.6%

9.34 (a) 0.186 (b) third place to fifth place are separated by less than 0.05, as are sixth and seventh (c) different sprinters have different strengths

9.35 (a) 0.801 (b) 0.856 (c) 0.848 (d) Ten yards is slightly more valuable in this situation.

Chapter 10: Big Data and Beyond

10.1 Suppose the opponents all have high catch-and-shoot percentages, low pull-up shooting percentages, and the point guard has a low shooting percentage on drives. The defense should stay close to potential shooters, trying to force dribbles after the pass or allowing the driver to shoot.

10.2 A player who is measurably slowing down as the match continues can be subbed out, or an opponent who was slowing down would not have to be marked as closely.

10.3 A player could see that, for example, the opponent always served to the backhand and slide over a step, or see that the opponent was returning serves from well behind the baseline and come in to the net.

10.4 The coach could see if a player was tired and slowing down, or injured and not cutting sharply, or fully recovered from an injury and moving well.

10.5 (a) the approach shot is $+0.36$ strokes worse than average, the putts -0.28 strokes better than average; the overall Strokes Gained is still -0.114, the difference between the actual score of 4 and the expected score. (b) the drive is -0.234, the approach shot $+0.28$, and the putts $+0.12$, for an overall $+0.166$ strokes.

10.6 (a) $f(1.98) = 31.1$ (b) $g(1.3) = 30.0$ (c) $h(3.4) = 17.0$ (d) largest effect of 7 percentage points for running time, then 6.286 for delivery time, then 2.75 for pop time (each equal to 0.1 times the coefficient of x in the appropriate formula)

10.7 (a) $241/70 = 3.44$ s (b) $f(1.8) = 36$ and $g(1.2) = 36.3$ so an estimate of 36% is reasonable (c) Pop time and delivery time are better than average, so the running time needs to be smaller; 3.44 is too high

10.8 (a) 250 (b) 2500 (c) 2500 extra calls is not realistic

10.9 (a) $5.8409/0.1078 = 54$ yards; looks reasonable (b) $(\ln 3 + 5.8409)/0.1078 = 64$ yards; still reasonable (c) $(\ln 9 + 5.8409)/0.1078 = 75$ yards; not likely

10.10 Kickers who stay employed are so accurate that a small number of misses separates the best from the worst, so bad luck can have a large effect.

10.11 (a) $y = -1$ (b) slope 1, point (5,0): $y = x - 5$ (c) slope $-1/2$, point (5,3): $y = -0.5(x - 5) + 3$

10.12 The dividing lines $y = x - 5$ (see 10.11(b)) and $y = 5 - x$ intersect at (5,0). Keep the portions of these lines to the right of $x = 5$ and the portion of the line $y = -1$ (see 10.11(a)) to the left of $(5, 0)$.

10.13 (a) A is closest: AP=4.1, BP=4.5, CP=4.2 (b) C is closest: AP=5.4, BP=4.5, CP=4.2

10.14 The convex hull is still the triangle ABC, which now has sides $\sqrt{10}$, $\sqrt{26}$, and $\sqrt{20}$. The area is 7. C is farther from A and B, the defense is more spread out, and the area is larger.

10.15 The convex hull is the triangle ABD, which now has sides $\sqrt{13}$, $\sqrt{2}$, and $\sqrt{17}$. The area is 2.5.

10.18 (a) 0.580 (b) 0.581

10.19 At 13-4, B would still be larger so B would need to lose two more games to finish 12-5.

10.20 (a) Playing out of conference means that A (or B) is not the opponent. Weaker teams could potentially schedule more easy wins and improve the C-OC mark. (b) The totals become 155-133 for A and 162-127 for B. B's advantage has decreased from 7 games to 6.5 games.

10.21 The reliability is increased by the extra data. A loss of charm could be overcome by these stats being made readily available on smart phones and video screens at games.

10.23 Receivers have time to adjust on long passes and can outmaneuver defenders. There is little time for a receiver to adjust on a short pass.

10.25 Neither addresses the quality of the pitcher's pick-off move, but slow delivery and pop times greatly increase the probability of a stolen base being successful.

10.26 Good framing technique has little motion, as if the pitcher is throwing to the exact intended spot. Blatant movement of the glove after catching the ball does not work.

10.27 The differences among kickers are small, so the performance differences could be largely a matter of luck.

10.28 All rating systems have flaws and blind spots, but this obvious bias is not good for a system used for important decisions.

10.29 It does not appear easier to rebound a 3-point shot as opposed to a 2-point shot, but this is a different question than long shot versus short shot.

10.30 (a) The denominator has derivative $(1 + e^{-5.8409+0.1078d})' = 0.1078e^{-5.8409+0.1078d} > 0$; since the denominator is increasing, the function is decreasing. The longer the kick, the lower the probability of making the kick. (b) $\frac{1}{1+e^{-5.8409}} = 0.997$ (c) 0 (d) the limit as d goes to 0 should be 1 (0.997 is close)

10.31 $\int_{-\pi/2}^{\pi/2} \int_1^{15} \frac{k}{r} r \, \mathrm{d}r \, \mathrm{d}\theta = 14k\pi$ so $k = \frac{1}{14\pi}$. A: 0.42, B: 0.58

10.32 (a) The trapezoid has height 2 and parallel sides of length 5 and 3, so its area is 8. (b) The areas are 3 for ABC and BCD and 5 for ABD and ACD. Adding up the areas double counts each region so half the sum gives a total area of 8.

10.33 (a) 0.163 (b) 0.005 (c) 0.186

10.34 (a) 0.108 (b) no, 2-2 has a probability of 0.122

Index